Georgia Milestones Assessment System Subject Test
Mathematics Grade 6

Student Practice Workbook

+ Two Full-Length GMAS Math Tests

Math Notion

www.MathNotion.com

GMAS Subject Test Mathematics Grade 6

GMAS Subject Test Mathematics Grade 6

Georgia Milestones Assessment System Subject Test

Mathematics Grade 6

Published in the United State of America By

The Math Notion

Web: WWW.MathNotion.com

Email: info@Mathnotion.com

Copyright © 2021 by the Math Notion. All rights reserved. No part of this publication may be reproduced, stored in a retrieval system, or transmitted in any form or by any means, electronic, mechanical, photocopying, recording, scanning, or otherwise, except as permitted under Section 107 or 108 of the 1976 United States Copyright Ac, without permission of the author.

All inquiries should be addressed to the Math Notion.

ISBN: 978-1-63620-074-3

GMAS Subject Test Mathematics Grade 6

The Math Notion

Michael Smith has been a math instructor for over a decade now. He launched the Math Notion. Since 2006, we have devoted our time to both teaching and developing exceptional math learning materials. As a test prep company, we have worked with thousands of students. We have used the feedback of our students to develop a unique study program that can be used by students to drastically improve their math scores fast and effectively. We have more than a thousand Math learning books including:

– **SAT Math Prep**

– **ACT Math Prep**

– **SSAT/ISEE Math Prep**

– **Accuplacer Math Prep**

– **Common Core Math Prep**

–many **Math Education Workbooks, Study Guides, Practice and Exercise Books**

As an experienced Math test preparation company, we have helped many students raise their standardized test scores—and attend the colleges of their dreams: We tutor online and in person, we teach students in large groups, and we provide training materials and textbooks through our website and through Amazon.

You can contact us via email at:

info@Mathnotion.com

GMAS Subject Test Mathematics Grade 6

Get the Targeted Practice You Need to Ace the GMAS Math Test!

Georgia Milestones Assessment System Subject Test Mathematics Grade 6 includes easy-to-follow instructions, helpful examples, and plenty of math practice problems to assist students to master each concept, brush up their problem-solving skills, and create confidence.

The GMAS math practice book provides numerous opportunities to evaluate basic skills along with abundant remediation and intervention activities. It is a skill that permits you to quickly master intricate information and produce better leads in less time.

Students can boost their test-taking skills by taking the book's two practice GMAS Math exams. All test questions answered and explained in detail.

Important Features of the 6th grade GMAS Math Book:

- A **complete review** of GMAS math test topics,
- Over 2,500 practice problems covering all topics tested,
- The most important concepts you need to know,
- Clear and concise, easy-to-follow sections,
- Well designed for enhanced learning and interest,
- Hands-on experience with all question types,
- **2 full-length practice tests** with detailed answer explanations,
- Cost-Effective Pricing,

Powerful math exercises to help you avoid traps and pacing yourself to beat the Georgia Milestones test. Students will gain valuable experience and raise their confidence by taking 6th grade math practice tests, learning about test structure, and gaining a deeper understanding of what is tested on the Georgia standards math grade 6. If ever there was a book to respond to the pressure to increase students' test scores, this is it.

WWW.MathNotion.COM

... So Much More Online!

- ✓ FREE Math Lessons
- ✓ More Math Learning Books!
- ✓ Mathematics Worksheets
- ✓ Online Math Tutors

For a PDF Version of This Book

Please Visit WWW.MathNotion.com

GMAS Subject Test Mathematics Grade 6

Contents

Chapter 1 : Review of the Whole Number Operations..................11
   Adding Whole Numbers 12
   Subtracting Whole Numbers 13
   Multiplying Whole Numbers...14
   Dividing Hundreds...15
   Long Division by Two Digits 16
   Division with Remainders 16
   Rounding Whole Numbers 17
   Whole Number Estimation 18
   Answers of Worksheets 19

Chapter 2 : Integers and Number Theory 21
   Adding and Subtracting Integers 22
   Multiplying and Dividing Integers 23
   Order of Operations..24
   Ordering Integers and Numbers..25
   Integers and Absolute Value 26
   Factoring Numbers 27
   Prime Factorization...27
   Divisibility Rules..28
   Greatest Common Factor..29
   Least Common Multiple 30
   Answers of Worksheets 31

Chapter 3 : Fractions 35
   Simplifying Fractions..36
   Adding and Subtracting Fractions 37
   Multiplying and Dividing Fractions 38
   Adding and Subtracting Mixed Numbers 39
   Multiplying and Dividing Mixed Numbers 40
   Answers of Worksheets 41

Chapter 4 : Decimals 43
   Adding and Subtracting Decimals...44
   Multiplying and Dividing Decimals...45
   Comparing Decimals..46

GMAS Subject Test Mathematics Grade 6

  Rounding Decimals 47
  Convert Fraction to Decimal 48
  Convert Decimal to Percent...49
  Convert Fraction to Percent 50
  Answers of Worksheets 51

Chapter 5 : Proportions, Ratios, and Percent..53
  Simplifying Ratios..54
  Proportional Ratios 55
  Similarity and Ratios 56
  Ratio and Rates Word Problems 57
  Percentage Calculations..58
  Percent Problems 59
  Discount, Tax and Tip 60
  Answers of Worksheets 61

Chapter 6 : Exponents and Radicals Expressions...63
  Adding and Subtracting Exponents 64
  Multiplication Property of Exponents 65
  Zero and Negative Exponents...66
  Division Property of Exponents 67
  Powers of Products and Quotients 68
  Negative Exponents and Negative Bases 69
  Scientific Notation 70
  Square Roots 71
  Answers of Worksheets 72

Chapter 7 : Measurements..75
  Reference Measurement..76
  Metric Length Measurement 77
  Customary Length Measurement..77
  Metric Capacity Measurement 78
  Customary Capacity Measurement 78
  Metric Weight and Mass Measurement 79
  Customary Weight and Mass Measurement 79
  Temperature 80
  Time 81
  Answers of Worksheets 82

Chapter 8 : Algebraic Expressions 84
  Find a Rule!..85

GMAS Subject Test Mathematics Grade 6

    Translate Phrases into an Algebraic Statement 86

    Simplifying Variable Expressions 87

    The Distributive Property 88

    Evaluating One Variable Expressions 89

    Combining like Terms 90

    Answers of Worksheets 91

Chapter 9 : Equations and Inequalities 93

    One–Step Equations 94

    One–Step Equation Word Problems 95

    Two-Steps Equations 96

    Multi–Step Equations 97

    One-Step Inequalities 98

    Graphing Inequalities 99

    Two-Steps Inequality 100

    Multi-Step Inequalities 101

    Answers of Worksheets 102

Chapter 10 : Geometry and Solid Figures 105

    Angles 106

    Pythagorean Relationship 107

    Triangles 108

    Polygons 109

    Trapezoids 110

    Circles 111

    Cubes 112

    Rectangular Prism 113

    Cylinder 114

    Answers of Worksheets 115

Chapter 11 : Statistics and Probability 117

    Mean and Median 118

    Mode and Range 119

    Times Series 120

    Stem–and–Leaf Plot 121

    Quartile of a Data Set 122

    Box and Whisker Plots 122

    Pie Graph 123

    Probability Problems 124

    Answers of Worksheets 125

WWW.MathNotion.Com

GMAS Subject Test Mathematics Grade 6

Chapter 12 : GMAS Math Practice Tests 127
  GMAS Grade 6 Mathematics Reference Sheet 129
  Georgia Milestones Assessment System Practice Test 1 131
    Session 1 132
    Session 2 136
  Georgia Milestones Assessment System Practice Test 2 141
    Session 1 142
    Session 2 146

Chapter 13 : Answers and Explanations 151
  Answer Key 151
  Practice Test 1 153
  Practice Test 2 157

GMAS Subject Test Mathematics Grade 6

Chapter 1 : Review of the Whole Number Operations

Topics that you'll learn in this chapter:

- ✓ Adding Whole Numbers
- ✓ Subtracting Whole Numbers
- ✓ Multiplying Whole Numbers
- ✓ Dividing Hundreds
- ✓ Long Division by One Digit
- ✓ Division with Remainders
- ✓ Rounding Whole Numbers
- ✓ Whole Number Estimation

"Wherever there is number, there is beauty." –Proclus

Adding Whole Numbers

✎ Add.

1) 5,763 + 8,238 = _____

2) 6,834 + 4,998 = _____

3) 3,548 + 5,693 = _____

4) 2,769 + 8,872 = _____

5) 3,196 + 2,936 = _____

6) 7,009 + 4,992 = _____

✎ Find the missing numbers.

7) 3,468 + ___ = 4,102

8) 840 + 2,360 = ___

9) 5,200 + ___ = 7,980

10) 631 + ___ = 2,007

11) ___ + 803 = 3,945

12) ___ + 2,156 = 5,922

13) David sells gems. He finds a diamond in Istanbul and buys it for $4,795. Then, he flies to Cairo and purchases a bigger diamond for the bargain price of $9,633. How much does David spend on the two diamonds? _____

GMAS Subject Test Mathematics Grade 6

Subtracting Whole Numbers

✏ **Subtract.**

1) 10,512
   −4,411
   _____

4) 8,001
   − 5,224
   _____

2) 5,204
   − 3,679
   _____

5) 11,916
   − 8,711
   _____

3) 8,520
   − 6,483
   _____

6) 5,005
   −2,008
   _____

✏ **Find the missing number.**

7) 5,263 − ___ = 2,367

10) 6,511 − ___ = 3,759

8) 7,198 − ___ = 4,742

11) 7,003 − 5,489 = ___

9) 8,928 − 3,764 = ___

12) 8,800 − 5,995 = ___

13) Jackson had $7,189 invested in the stock market until he lost $3,793 on those investments. How much money does he have in the stock market now?

_____

WWW.MathNotion.Com

GMAS Subject Test Mathematics Grade 6

Multiplying Whole Numbers

✏ **Find the answers.**

1) 2,200 × 31 = _____

2) 3,200 × 22 = _____

3) 5,790 × 5 = _____

4) 5,220 × 3 = _____

5) 6,911 × 3 = _____

6) 1,998 × 40 = _____

7) 2,893 × 5.5 = _____

8) 2,254 × 3.5 = _____

9) 4,372 × 4.8 = _____

10) 3,984 × 2.75 = _____

11) 4,900 × 2.5 = _____

12) 8,200 × 4.5 = _____

GMAS Subject Test Mathematics Grade 6

Dividing Hundreds

✎ **Find answers.**

1) $4,440 \div 400$

2) $1,600 \div 40$

3) $9,990 \div 90$

4) $4,200 \div 60$

5) $6,400 \div 8,000$

6) $2,700 \div 30$

7) $3,333 \div 30$

8) $558 \div 45$

9) $2,278 \div 85$

10) $1,683 \div 55$

11) $1,582 \div 35$

12) $9,000 \div 600$

13) $1,000 \div 2,500$

14) $44.8 \div 20$

15) $6,800 \div 400$

16) $1,500 \div 5,000$

17) $36.60 \div 120$

18) $7,700 \div 700$

19) $5,400 \div 600$

20) $8,000 \div 160$

21) $18,000 \div 9,000$

22) $42,000 \div 30$

23) $480 \div 40$

24) $63,000 \div 900$

WWW.MathNotion.Com

GMAS Subject Test Mathematics Grade 6

Long Division by Two Digits

✏️ **Find the quotient.**

1) $18\overline{)576}$

2) $14\overline{)952}$

3) $21\overline{)588}$

4) $23\overline{)299}$

5) $44\overline{)748}$

6) $26\overline{)234}$

7) $16\overline{)496}$

8) $29\overline{)1,479}$

9) $54\overline{)1,080}$

10) $41\overline{)1,476}$

11) $53\overline{)2,491}$

12) $60\overline{)2,880}$

13) $32\overline{)2,912}$

14) $77\overline{)8,393}$

15) $85\overline{)3,740}$

16) $57\overline{)4,617}$

17) $50\overline{)9,200}$

18) $25\overline{)15,400}$

Division with Remainders

✏️ **Find the quotient with remainder.**

1) $14\overline{)715}$

2) $16\overline{)2,750}$

3) $27\overline{)4,603}$

4) $58\overline{)2,554}$

5) $42\overline{)7,732}$

6) $63\overline{)6,737}$

7) $71\overline{)9,036}$

8) $65\overline{)8,624}$

9) $35\overline{)5,705}$

10) $92\overline{)13,161}$

11) $46\overline{)12,214}$

12) $69\overline{)42,482}$

13) $85\overline{)6,858}$

14) $87\overline{)34,304}$

GMAS Subject Test Mathematics Grade 6

Rounding Whole Numbers

✎ **Round each number to the underlined place value.**

1) 7,<u>5</u>33

2) 9,<u>3</u>74

3) 8,8<u>8</u>3

4) 2,3<u>6</u>8

5) 5,5<u>7</u>7

6) 3,3<u>8</u>1

7) 3,<u>5</u>20

8) 9,3<u>3</u>8

9) 8.<u>5</u>81

10) 33.<u>5</u>7

11) 51.<u>6</u>9

12) 22.<u>1</u>38

13) <u>6</u>,758

14) 11,<u>5</u>57

15) 8,8<u>3</u>8

16) 5.<u>8</u>89

17) 1.<u>8</u>60

18) 25.<u>0</u>70

19) <u>9</u>.332

20) 49.<u>4</u>8

21) 28.<u>8</u>9

22) 24,3<u>7</u>7

23) 52,1<u>5</u>8

24) 13,8<u>8</u>3

25) 9,<u>6</u>09

26) 17,4<u>5</u>1

27) 18,<u>7</u>68

WWW.MathNotion.Com

GMAS Subject Test Mathematics Grade 6

Whole Number Estimation

✏️ **Estimate the sum by rounding each added to the nearest ten.**

1) 875 + 325

2) 985 + 1,452

3) 2,424 + 4,128

4) 1,576 + 6,279

5) 1,247 + 3,863

6) 6,746 + 5,121

7) 3,924 + 6,456

8) 1,785 + 7,164

9) 1,458
   + 2,442
   ─────

10) 5,689
    + 4,151
    ─────

11) 8,259
    + 4,754
    ─────

12) 6,788
    + 3,954
    ─────

13) 9,123
    + 4,455
    ─────

14) 6,680
    + 5,358
    ─────

15) 3,165
    + 7,124
    ─────

16) 8,859
    + 6,452
    ─────

WWW.MathNotion.Com

GMAS Subject Test Mathematics Grade 6

Answers of Worksheets

Adding Whole Numbers

1) 14,001
2) 11,832
3) 9,241
4) 11,641
5) 6,132
6) 12,001
7) 634
8) 3,200
9) 2,780
10) 1,376
11) 3,142
12) 3,766
13) $14,428

Subtracting Whole Numbers

1) 6,101
2) 1,525
3) 2,037
4) 2,777
5) 3,205
6) 2,997
7) 2,896
8) 2,456
9) 5,164
10) 2,752
11) 1,514
12) 2,805
13) 3,396

Multiplying Whole Numbers

1) 68,200
2) 70,400
3) 28,950
4) 15,660
5) 20,733
6) 79,920
7) 15,911.5
8) 7,889
9) 20,985.6
10) 10,956
11) 12,250
12) 36,900

Dividing Hundreds

1) 11.1
2) 40
3) 111
4) 70
5) 0.8
6) 90
7) 111.1
8) 12.4
9) 26.8
10) 30.6
11) 45.2
12) 15
13) 0.4
14) 2.24
15) 17
16) 0.3
17) 0.305
18) 11
19) 9
20) 50
21) 2
22) 1,400
23) 12
24) 70

Long Division by Two Digits

1) 32
2) 68
3) 28
4) 13
5) 17
6) 9
7) 31
8) 51
9) 20
10) 36
11) 47
12) 48
13) 91
14) 109
15) 44
16) 81
17) 184
18) 616

GMAS Subject Test Mathematics Grade 6

Division with Remainders

1) 51 R1
2) 171 R14
3) 170 R13
4) 44 R2
5) 184 R4
6) 106 R59
7) 127 R19
8) 132 R44
9) 163 R0
10) 143 R5
11) 265 R24
12) 615 R47
13) 80 R58
14) 394 R26

Rounding Whole Numbers

1) 7,500
2) 9,400
3) 8,880
4) 2,370
5) 5,580
6) 3,380
7) 3,500
8) 9,340
9) 8.60
10) 33.60
11) 51.70
12) 22.100
13) 7,000
14) 11,560
15) 8,840
16) 5.900
17) 1.900
18) 25.100
19) 9.000
20) 49.50
21) 28.90
22) 24,380
23) 52,160
24) 13,880
25) 9,600
26) 17,450
27) 18,800

Whole Number Estimation

1) 1,200
2) 2,440
3) 6,550
4) 7,860
5) 5,110
6) 11,870
7) 10,380
8) 8,950
9) 3,900
10) 9,840
11) 13,010
12) 10,740
13) 13,580
14) 12,040
15) 10,290
16) 15,310

GMAS Subject Test Mathematics Grade 6

Chapter 2 :
Integers and Number Theory

Topics that you will practice in this chapter:

- ✓ Adding and Subtracting Integers
- ✓ Multiplying and Dividing Integers
- ✓ Order of Operations
- ✓ Ordering Integers and Numbers
- ✓ Integers and Absolute Value
- ✓ Factoring Numbers
- ✓ Prime Factorization
- ✓ Divisibility Rules
- ✓ Greatest Common Factor (GCF)
- ✓ Least Common Multiple (LCM)

"In order to gain the most, you have to know how to convert Negatives to Positives."

—*Stubborn Clown*

Adding and Subtracting Integers

✍ **Find each sum.**

1) $14 + (-6) =$

2) $(-13) + (-20) =$

3) $5 + (-28) =$

4) $50 + (-12) =$

5) $(-7) + (-15) + 3 =$

6) $30 + (-14) + 8 =$

7) $40 + (-10) + (-14) + 17 =$

8) $(-15) + (-20) + 13 + 35 =$

9) $40 + (-20) + (38 - 29) =$

10) $28 + (-12) + (30 - 12) =$

✍ **Find each difference.**

11) $(-18) - (-7) =$

12) $25 - (-14) =$

13) $(-20) - 36 =$

14) $34 - (-19) =$

15) $51 - (30 - 21) =$

16) $17 - (5) - (-24) =$

17) $(35 + 20) - (-46) =$

18) $48 - 16 - (-8) =$

19) $62 - (28 + 17) - (-15) =$

20) $58 - (-23) - (-31) =$

21) $19 - (-8) - (-13) =$

22) $(19 - 24) - (-14) =$

23) $27 - 33 - (-21) =$

24) $58 - (32 + 24) - (-9) =$

25) $36 - (-30) + (-17) =$

26) $27 - (-42) + (-31) =$

GMAS Subject Test Mathematics Grade 6

Multiplying and Dividing Integers

✏️ **Find each product.**

1) $(-9) \times (-5) =$

2) $(-3) \times 9 =$

3) $8 \times (-12) =$

4) $(-7) \times (-20) =$

5) $(-3) \times (-5) \times 6 =$

6) $(14 - 3) \times (-8) =$

7) $12 \times (-9) \times (-3) =$

8) $(140 + 10) \times (-2) =$

9) $10 \times (-12 + 8) \times 3 =$

10) $(-8) \times (-5) \times (-10) =$

✏️ **Find each quotient.**

11) $42 \div (-7) =$

12) $(-48) \div (-6) =$

13) $(-40) \div (-8) =$

14) $54 \div (-2) =$

15) $152 \div 19 =$

16) $(-144) \div (-12) =$

17) $180 \div (-10) =$

18) $(-312) \div (-12) =$

19) $221 \div (-13) =$

20) $(-126) \div (6) =$

21) $(-161) \div (-7) =$

22) $-266 \div (-14) =$

23) $(-120) \div (-4) =$

24) $270 \div (-18) =$

25) $(-208) \div (-8) =$

26) $(135) \div (-15) =$

WWW.MathNotion.Com

GMAS Subject Test Mathematics Grade 6

Order of Operations

✎ Evaluate each expression.

1) $7 + (5 \times 4) =$

2) $14 - (3 \times 6) =$

3) $(19 \times 4) + 16 =$

4) $(16 - 7) - (8 \times 2) =$

5) $27 + (18 \div 3) =$

6) $(18 \times 8) \div 6 =$

7) $(32 \div 4) \times (-2) =$

8) $(9 \times 4) + (32 - 18) =$

9) $24 + (4 \times 3) + 7 =$

10) $(36 \times 3) \div (2 + 2) =$

11) $(-7) + (12 \times 3) + 11 =$

12) $(8 \times 5) - (24 \div 6) =$

13) $(7 \times 6 \div 3) - (12 + 9) =$

14) $(13 + 5 - 14) \times 3 - 2 =$

15) $(20 - 14 + 30) \times (64 \div 4) =$

16) $32 + (28 - (36 \div 9)) =$

17) $(7 + 6 - 4 - 7) + (15 \div 5) =$

18) $(85 - 20) + (20 - 18 + 7) =$

19) $(20 \times 2) + (14 \times 3) - 22 =$

20) $18 + 5 - (30 \times 3) + 20 =$

21) $(\frac{7}{5-1}) \times (2 + 6) \times 2$

22) $20 \div (4 - (10 - 8))$

WWW.MathNotion.Com

GMAS Subject Test Mathematics Grade 6

Ordering Integers and Numbers

✎ **Order each set of integers from least to greatest.**

1) $8, -10, -5, -3, 4$ ___, ___, ___, ___, ___, ___

2) $-10, -18, 6, 14, 27$ ___, ___, ___, ___, ___, ___

3) $15, -8, -21, 21, -23$ ___, ___, ___, ___, ___, ___

4) $-14, -40, 23, -12, 47$ ___, ___, ___, ___, ___, ___

5) $59, -54, 32, -57, 36$ ___, ___, ___, ___, ___, ___

6) $68, 26, -19, 47, -34$ ___, ___, ___, ___, ___, ___

✎ **Order each set of integers from greatest to least.**

7) $18, 36, -16, -18, -10$ ___, ___, ___, ___, ___, ___

8) $27, 34, -12, -24, 94$ ___, ___, ___, ___, ___, ___

9) $50, -21, -13, 42, -2$ ___, ___, ___, ___, ___, ___

10) $37, 46, -20, -16, 86$ ___, ___, ___, ___, ___, ___

11) $-18, 88, -26, -59, 75$ ___, ___, ___, ___, ___, ___

12) $-65, -30, -25, 3, 14$ ___, ___, ___, ___, ___, ___

GMAS Subject Test Mathematics Grade 6

Integers and Absolute Value

✎ **Write absolute value of each number.**

1) $|-2| =$

2) $|-27| =$

3) $|-20| =$

4) $|14| =$

5) $|6| =$

6) $|-55| =$

7) $|16| =$

8) $|2| =$

9) $|54| =$

10) $|-4| =$

11) $|-11|$

12) $|88| =$

13) $|0| =$

14) $|79| =$

15) $|-32| =$

16) $|-17| =$

17) $|42| =$

18) $|-46| =$

19) $|1| =$

20) $|-40| =$

✎ **Evaluate the value.**

21) $|-5| - \frac{|-21|}{7} =$

22) $14 - |3 - 15| - |-4| =$

23) $\frac{|-32|}{4} \times |-4| =$

24) $\frac{|7 \times (-3)|}{7} \times \frac{|-19|}{3} =$

25) $|4 \times (-5)| + \frac{|-40|}{5} =$

26) $\frac{|-45|}{9} \times \frac{|-24|}{12} =$

27) $|-12 + 8| \times \frac{|-7 \times 7|}{7}$

28) $\frac{|-11 \times 2|}{4} \times |-16| =$

WWW.MathNotion.Com

GMAS Subject Test Mathematics Grade 6

Factoring Numbers

✎ List all positive factors of each number.

1) 12
2) 16
3) 28
4) 34
5) 95
6) 56
7) 65
8) 70
9) 25
10) 48
11) 27
12) 63
13) 72
14) 15
15) 80

✎ List the prime factorization for each number.

16) 10
17) 26
18) 20
19) 30
20) 40
21) 44
22) 55
23) 78
24) 96

Prime Factorization

✎ Factor the following numbers to their prime factors.

1) 6
2) 49
3) 60
4) 4
5) 46
6) 57
7) 54
8) 38
9) 58
10) 62
11) 75
12) 88
13) 93
14) 100
15) 68
16) 90
17) 69
18) 76
19) 86
20) 92
21) 99
22) 77
23) 90
24) 74

Divisibility Rules

✎ **Use the divisibility rules to underline the factors of the number.**

1) 8   2 3 4 5 6 7 8 9 10

2) 18   2 3 4 5 6 7 8 9 10

3) 55   2 3 4 5 6 7 8 9 10

4) 45   2 3 4 5 6 7 8 9 10

5) 20   2 3 4 5 6 7 8 9 10

6) 9   2 3 4 5 6 7 8 9 10

7) 21   2 3 4 5 6 7 8 9 10

8) 28   2 3 4 5 6 7 8 9 10

9) 36   2 3 4 5 6 7 8 9 10

10) 40 2 3 4 5 6 7 8 9 10

11) 39 2 3 4 5 6 7 8 9 10

12) 51 2 3 4 5 6 7 8 9 10

Greatest Common Factor

✎ **Find the GCF for each number pair.**

1) 6, 2

2) 4, 5

3) 3, 12

4) 7, 3

5) 5, 10

6) 8, 48

7) 6, 18

8) 9, 15

9) 12, 18

10) 4, 36

11) 6, 10

12) 28, 52

13) 25, 10

14) 22, 24

15) 9, 54

16) 8, 54

17) 42, 14

18) 16, 40

19) 9, 2, 3

20) 5, 15, 10

21) 7, 9, 2

22) 16, 64

23) 30, 48

24) 36, 63

GMAS Subject Test Mathematics Grade 6

Least Common Multiple

✎ Find the LCM for each number pair.

1) 6, 9

2) 15, 45

3) 16, 40

4) 12, 36

5) 18, 27

6) 14, 42

7) 6, 30

8) 8, 56

9) 7, 21

10) 8, 20

11) 15, 25

12) 7, 9

13) 4, 11

14) 8, 28

15) 28, 56

16) 40, 50

17) 12, 13

18) 22, 11

19) 36, 20

20) 15, 35

21) 18, 81

22) 30, 54

23) 18, 45

24) 75, 25

WWW.MathNotion.Com

GMAS Subject Test Mathematics Grade 6

Answers of Worksheets

Adding and Subtracting Integers

1) 8	8) 13	15) 42	22) 9
2) −33	9) 29	16) 36	23) 15
3) −23	10) 34	17) 101	24) 11
4) 38	11) −11	18) 40	25) 49
5) −19	12) 39	19) 32	26) 38
6) 24	13) −56	20) 112	
7) 33	14) 53	21) 40	

Multiplying and Dividing Integers

1) 45	8) −300	15) 8	22) 19
2) −27	9) −120	16) 12	23) 30
3) −96	10) −400	17) −18	24) −15
4) 140	11) −6	18) 26	25) 26
5) 90	12) 8	19) −17	26) −9
6) −88	13) 5	20) −21	
7) 324	14) −27	21) 23	

Order of Operations

1) 27	7) −16	13) −7	19) 60
2) −4	8) 50	14) 10	20) −47
3) 92	9) 43	15) 576	21) 28
4) −7	10) 27	16) 56	22) 10
5) 33	11) 40	17) 5	
6) 24	12) 36	18) 74	

Ordering Integers and Numbers

1) −10, −5, −3, 4, 8
2) −18, −10, 6, 14, 27
3) −23, −21, −8, 15, 21
4) −40, −14, −12, 23, 47
5) −57, −54, 32, 36, 59
6) −34, −19, 26, 47, 68
7) 36, 18, −10, −16, −18
8) 94, 34, 27, −12, −24
9) 50, 42, −2, −13, −21
10) 86, 46, 37, −16, −20
11) 88, 75, −18, −26, −59
12) 14, 3, −25, −30, −65

WWW.MathNotion.Com

GMAS Subject Test Mathematics Grade 6

Integers and Absolute Value

1) 2	8) 2	15) 32	22) −2
2) 27	9) 54	16) 17	23) 32
3) 20	10) 4	17) 42	24) 19
4) 14	11) 11	18) 46	25) 28
5) 6	12) 88	19) 1	26) 10
6) 55	13) 0	20) 40	27) 28
7) 16	14) 79	21) 2	28) 88

Factoring Numbers

1) 1, 2, 3, 4, 6, 12
2) 1, 2, 4, 8, 16
3) 1, 2, 4, 7, 14, 28
4) 1, 2, 17, 34
5) 1, 5, 19, 95
6) 1, 2, 4, 7, 8, 14, 28, 56
7) 1, 5, 13, 65
8) 1, 2, 5, 7, 10, 14, 35, 70
9) 1, 5, 25
10) 1, 2, 3, 4, 6, 8, 12, 16, 24, 48
11) 1, 3, 9, 27
12) 1, 3, 7, 9, 21, 63
13) 1, 2, 3, 4, 6, 8, 9, 12, 18, 24, 36, 72
14) 1, 3, 5, 15
15) 1, 2, 4, 5, 8, 10, 16, 20, 40, 80
16) 2 × 5
17) 2 × 13
18) 2 × 2 × 5
19) 2 × 3 × 5
20) 2 × 2 × 2 × 5
21) 2 × 2 × 11
22) 5 × 11
23) 2 × 3 × 13
24) 2 × 2 × 2 × 2 × 2 × 3

Prime Factorization

1) 2. 3	9) 2. 29	17) 3. 23
2) 7. 7	10) 2. 31	18) 2. 2. 19
3) 2. 2. 3. 5	11) 3. 5. 5	19) 2. 43
4) 2. 2	12) 2. 2. 2. 11	20) 2. 2. 23
5) 2. 23	13) 3. 31	21) 3. 3. 11
6) 3. 19	14) 2. 2. 5. 5	22) 7. 11
7) 2. 3. 3. 3	15) 2. 2. 17	23) 2. 3. 3. 5
8) 2. 19	16) 2. 3. 3. 5	24) 2. 37

GMAS Subject Test Mathematics Grade 6

Divisibility Rules

1) 8   <u>2</u> 3 <u>4</u> 5 6 7 <u>8</u> 9 10
2) 18   <u>2</u> <u>3</u> 4 5 <u>6</u> 7 8 <u>9</u> 10
3) 55   2 3 4 <u>5</u> 6 7 8 9 10
4) 45   2 <u>3</u> 4 <u>5</u> 6 7 8 <u>9</u> 10
5) 20   <u>2</u> 3 <u>4</u> <u>5</u> 6 7 8 9 <u>10</u>
6) 9   2 <u>3</u> 4 5 6 7 8 <u>9</u> 10
7) 21   2 <u>3</u> 4 5 6 <u>7</u> 8 9 10
8) 28   <u>2</u> 3 <u>4</u> 5 6 <u>7</u> 8 9 10
9) 36   <u>2</u> <u>3</u> <u>4</u> 5 <u>6</u> 7 8 <u>9</u> 10
10) 40 <u>2</u> 3 <u>4</u> <u>5</u> 6 7 <u>8</u> 9 <u>10</u>
11) 39 2 <u>3</u> 4 5 6 7 8 9 10
12) 51 2 <u>3</u> 4 5 6 7 8 9 10

Greatest Common Factor

1) 2
2) 1
3) 3
4) 1
5) 5
6) 8
7) 6
8) 3
9) 6
10) 4
11) 2
12) 4
13) 5
14) 2
15) 9
16) 2
17) 14
18) 8
19) 1
20) 5
21) 1
22) 16
23) 6
24) 9

Least Common Multiple

1) 18
2) 45
3) 80
4) 36
5) 54
6) 42
7) 30
8) 56
9) 21
10) 40
11) 75
12) 63
13) 44
14) 56
15) 56
16) 200
17) 156
18) 22
19) 180
20) 105
21) 162
22) 270
23) 90
24) 75

GMAS Subject Test Mathematics Grade 6

Chapter 3 :
Fractions

Topics that you will practice in this chapter:

- ✓ Simplifying Fractions
- ✓ Adding and Subtracting Fractions
- ✓ Multiplying and Dividing Fractions
- ✓ Adding and Subtract Mixed Numbers
- ✓ Multiplying and Dividing Mixed Numbers

"A Man is like a fraction whose numerator is what he is and whose denominator is what he thinks of himself. The larger the denominator, the smaller the fraction." –Tolstoy

GMAS Subject Test Mathematics Grade 6

Simplifying Fractions

✍ Simplify each fraction to its lowest terms.

1) $\dfrac{5}{10} =$

2) $\dfrac{28}{35} =$

3) $\dfrac{27}{36} =$

4) $\dfrac{40}{80} =$

5) $\dfrac{14}{56} =$

6) $\dfrac{32}{48} =$

7) $\dfrac{52}{65} =$

8) $\dfrac{15}{60} =$

9) $\dfrac{80}{160} =$

10) $\dfrac{55}{77} =$

11) $\dfrac{28}{112} =$

12) $\dfrac{32}{64} =$

13) $\dfrac{63}{72} =$

14) $\dfrac{81}{90} =$

15) $\dfrac{35}{105} =$

16) $\dfrac{25}{70} =$

17) $\dfrac{80}{280} =$

18) $\dfrac{12}{81} =$

19) $\dfrac{36}{186} =$

20) $\dfrac{240}{540} =$

21) $\dfrac{70}{560} =$

✍ Find the answer for each problem.

22) Which of the following fractions equal to $\dfrac{3}{4}$? ____

　A. $\dfrac{60}{90}$　　B. $\dfrac{43}{104}$　　C. $\dfrac{48}{64}$　　D. $\dfrac{150}{300}$

23) Which of the following fractions equal to $\dfrac{5}{8}$? ____

　A. $\dfrac{125}{200}$　　B. $\dfrac{115}{200}$　　C. $\dfrac{50}{100}$　　D. $\dfrac{30}{90}$

24) Which of the following fractions equal to $\dfrac{3}{7}$? ____

　A. $\dfrac{58}{116}$　　B. $\dfrac{54}{126}$　　C. $\dfrac{270}{167}$　　D. $\dfrac{42}{63}$

WWW.MathNotion.Com

GMAS Subject Test Mathematics Grade 6

Adding and Subtracting Fractions

✏️ **Find the sum.**

1) $\frac{5}{9} + \frac{4}{9} =$

2) $\frac{1}{2} + \frac{1}{7} =$

3) $\frac{3}{8} + \frac{1}{4} =$

4) $\frac{3}{5} + \frac{1}{2} =$

5) $\frac{1}{4} + \frac{3}{5} =$

6) $\frac{7}{8} + \frac{3}{8} =$

7) $\frac{1}{2} + \frac{7}{10} =$

8) $\frac{2}{5} + \frac{2}{3} =$

9) $\frac{5}{7} + \frac{2}{3} =$

10) $\frac{7}{12} + \frac{3}{4} =$

11) $\frac{5}{6} + \frac{2}{5} =$

12) $\frac{1}{12} + \frac{2}{3} =$

✏️ **Find the difference.**

13) $\frac{1}{3} - \frac{1}{6} =$

14) $\frac{3}{4} - \frac{1}{8} =$

15) $\frac{1}{2} - \frac{1}{3} =$

16) $\frac{1}{4} - \frac{1}{5} =$

17) $\frac{5}{8} - \frac{2}{3} =$

18) $\frac{1}{4} - \frac{1}{7} =$

19) $\frac{5}{6} - \frac{1}{9} =$

20) $\frac{3}{4} - \frac{1}{6} =$

21) $\frac{7}{8} - \frac{1}{12} =$

22) $\frac{8}{15} - \frac{3}{5} =$

23) $\frac{3}{12} - \frac{1}{14} =$

24) $\frac{10}{13} - \frac{7}{26} =$

25) $\frac{6}{7} - \frac{3}{4} =$

26) $\frac{4}{5} - \frac{1}{8} =$

27) $\frac{4}{7} - \frac{2}{35} =$

28) $\frac{9}{16} - \frac{2}{8} =$

29) $\frac{8}{9} - \frac{7}{18} =$

30) $\frac{1}{2} - \frac{4}{9} =$

GMAS Subject Test Mathematics Grade 6

Multiplying and Dividing Fractions

✏️ **Find the value of each expression in lowest terms.**

1) $\dfrac{1}{5} \times \dfrac{15}{5} =$

2) $\dfrac{9}{12} \times \dfrac{4}{9} =$

3) $\dfrac{1}{16} \times \dfrac{8}{10} =$

4) $\dfrac{1}{24} \times \dfrac{8}{10} =$

5) $\dfrac{1}{5} \times \dfrac{1}{4} =$

6) $\dfrac{7}{9} \times \dfrac{1}{7} =$

7) $\dfrac{6}{7} \times \dfrac{1}{3} =$

8) $\dfrac{2}{8} \times \dfrac{2}{8} =$

9) $\dfrac{5}{8} \times \dfrac{3}{5} =$

10) $\dfrac{4}{7} \times \dfrac{1}{8} =$

11) $\dfrac{7}{15} \times \dfrac{5}{7} =$

12) $\dfrac{3}{10} \times \dfrac{5}{9} =$

✏️ **Find the value of each expression in lowest terms.**

13) $\dfrac{1}{4} \div \dfrac{1}{8} =$

14) $\dfrac{1}{10} \div \dfrac{1}{5} =$

15) $\dfrac{3}{4} \div \dfrac{1}{5} =$

16) $\dfrac{1}{3} \div \dfrac{5}{6} =$

17) $\dfrac{1}{7} \div \dfrac{8}{42} =$

18) $\dfrac{3}{4} \div \dfrac{1}{6} =$

19) $\dfrac{2}{7} \div \dfrac{7}{13} =$

20) $\dfrac{1}{24} \div \dfrac{3}{16} =$

21) $\dfrac{7}{12} \div \dfrac{5}{6} =$

22) $\dfrac{22}{18} \div \dfrac{11}{9} =$

23) $\dfrac{9}{35} \div \dfrac{3}{7} =$

24) $\dfrac{2}{7} \div \dfrac{8}{21} =$

25) $\dfrac{1}{9} \div \dfrac{2}{5} =$

26) $\dfrac{5}{12} \div \dfrac{3}{5} =$

27) $\dfrac{3}{20} \div \dfrac{1}{6} =$

28) $\dfrac{8}{20} \div \dfrac{3}{4} =$

29) $\dfrac{5}{6} \div \dfrac{2}{9} =$

30) $\dfrac{5}{11} \div \dfrac{3}{4} =$

WWW.MathNotion.Com

GMAS Subject Test Mathematics Grade 6

Adding and Subtracting Mixed Numbers

✎ **Find the sum.**

1) $3\frac{1}{3} + 2\frac{1}{6} =$

2) $4\frac{1}{2} + 3\frac{1}{2} =$

3) $3\frac{3}{8} + 1\frac{1}{8} =$

4) $2\frac{1}{4} + 2\frac{1}{3} =$

5) $3\frac{5}{6} + 2\frac{7}{12} =$

6) $5\frac{4}{15} + 3\frac{3}{5} =$

7) $2\frac{1}{3} + 4\frac{3}{7} =$

8) $3\frac{1}{2} + 4\frac{2}{5} =$

9) $5\frac{2}{5} + 6\frac{3}{7} =$

10) $8\frac{5}{16} + 6\frac{1}{12} =$

✎ **Find the difference.**

11) $3\frac{1}{4} - 1\frac{3}{4} =$

12) $6\frac{3}{5} - 4\frac{2}{5} =$

13) $4\frac{1}{3} - 3\frac{1}{9} =$

14) $7\frac{1}{7} - 5\frac{1}{2} =$

15) $5\frac{1}{3} - 2\frac{1}{12} =$

16) $8\frac{1}{5} - 4\frac{1}{3} =$

17) $9\frac{1}{4} - 6\frac{1}{8} =$

18) $11\frac{7}{15} - 8\frac{3}{5} =$

19) $14\frac{5}{6} - 11\frac{3}{5} =$

20) $18\frac{2}{7} - 14\frac{1}{5} =$

21) $9\frac{1}{3} - 4\frac{1}{4} =$

22) $6\frac{1}{8} - 4\frac{1}{16} =$

23) $19\frac{3}{8} - 15\frac{1}{3} =$

24) $11\frac{1}{9} - 8\frac{1}{8} =$

25) $17\frac{1}{7} - 11\frac{1}{5} =$

26) $16\frac{2}{9} - 9\frac{5}{7} =$

WWW.MathNotion.Com

GMAS Subject Test Mathematics Grade 6

Multiplying and Dividing Mixed Numbers

✍ **Find the product.**

1) $5\frac{1}{2} \times 2\frac{1}{4} =$

2) $5\frac{1}{3} \times 4\frac{1}{3} =$

3) $5\frac{3}{4} \times 6\frac{1}{4} =$

4) $3\frac{1}{3} \times 2\frac{3}{5} =$

5) $4\frac{8}{10} \times 1\frac{1}{24} =$

6) $6\frac{2}{7} \times 1\frac{1}{11} =$

7) $8\frac{2}{3} \times 3\frac{1}{2} =$

8) $3\frac{4}{7} \times 2\frac{1}{5} =$

9) $5\frac{2}{8} \times 4\frac{1}{6} =$

10) $7\frac{3}{3} \times 1\frac{3}{8} =$

✍ **Find the quotient.**

11) $2\frac{2}{5} \div 4\frac{1}{5} =$

12) $4\frac{1}{6} \div 3\frac{1}{3} =$

13) $6\frac{1}{3} \div 1\frac{1}{2} =$

14) $7\frac{1}{10} \div 2\frac{2}{5} =$

15) $3\frac{1}{3} \div 1\frac{1}{9} =$

16) $1\frac{1}{10} \div 4\frac{1}{2} =$

17) $1\frac{3}{16} \div 5\frac{1}{4} =$

18) $4\frac{1}{3} \div 4\frac{3}{4} =$

19) $9\frac{1}{3} \div 2\frac{1}{4} =$

20) $15\frac{1}{3} \div 5\frac{1}{2} =$

21) $4\frac{1}{6} \div 1\frac{1}{5} =$

22) $1\frac{1}{18} \div 1\frac{2}{9} =$

23) $4\frac{2}{7} \div 1\frac{3}{10} =$

24) $7\frac{1}{3} \div 2\frac{2}{11} =$

25) $8\frac{2}{5} \div 1\frac{1}{6} =$

26) $9\frac{1}{3} \div 2\frac{1}{7} =$

GMAS Subject Test Mathematics Grade 6

Answers of Worksheets

Simplifying Fractions

1) $\frac{1}{2}$
2) $\frac{4}{5}$
3) $\frac{3}{4}$
4) $\frac{1}{2}$
5) $\frac{1}{4}$
6) $\frac{2}{3}$
7) $\frac{4}{5}$
8) $\frac{1}{4}$
9) $\frac{1}{2}$
10) $\frac{5}{7}$
11) $\frac{1}{4}$
12) $\frac{1}{2}$
13) $\frac{7}{8}$
14) $\frac{9}{10}$
15) $\frac{1}{3}$
16) $\frac{5}{14}$
17) $\frac{2}{7}$
18) $\frac{4}{27}$
19) $\frac{6}{31}$
20) $\frac{4}{9}$
21) $\frac{1}{8}$
22) C
23) A
24) B

Adding and Subtracting Fractions

1) $\frac{9}{9} = 1$
2) $\frac{9}{14}$
3) $\frac{5}{8}$
4) $1\frac{1}{10}$
5) $\frac{17}{20}$
6) $1\frac{1}{4}$
7) $1\frac{1}{5}$
8) $1\frac{1}{15}$
9) $1\frac{8}{21}$
10) $1\frac{1}{3}$
11) $1\frac{7}{30}$
12) $\frac{3}{4}$
13) $\frac{1}{6}$
14) $\frac{5}{8}$
15) $\frac{1}{6}$
16) $\frac{1}{20}$
17) $-\frac{1}{24}$
18) $\frac{3}{28}$
19) $\frac{13}{18}$
20) $\frac{7}{12}$
21) $\frac{19}{24}$
22) $-\frac{1}{15}$
23) $\frac{5}{28}$
24) $\frac{1}{2}$
25) $\frac{3}{28}$
26) $\frac{27}{40}$
27) $\frac{18}{35}$
28) $\frac{5}{16}$
29) $\frac{1}{2}$
30) $\frac{1}{18}$

Multiplying and Dividing Fractions

1) $\frac{3}{5}$
2) $\frac{1}{3}$
3) $\frac{1}{20}$
4) $\frac{1}{30}$
5) $\frac{1}{20}$
6) $\frac{1}{9}$
7) $\frac{2}{7}$
8) $\frac{1}{16}$
9) $\frac{3}{8}$
10) $\frac{1}{14}$
11) $\frac{1}{3}$
12) $\frac{1}{6}$
13) 2
14) $\frac{1}{2}$
15) $3\frac{3}{4}$
16) $\frac{2}{5}$

WWW.MathNotion.Com

GMAS Subject Test Mathematics Grade 6

17) $\frac{3}{4}$

18) $4\frac{1}{2}$

19) $\frac{26}{49}$

20) $\frac{2}{9}$

21) $\frac{7}{10}$

22) 1

23) $\frac{3}{5}$

24) $\frac{3}{4}$

25) $\frac{5}{18}$

26) $\frac{25}{36}$

27) $\frac{9}{10}$

28) $\frac{8}{15}$

29) $3\frac{3}{4}$

30) $\frac{20}{33}$

Adding and Subtracting Mixed Numbers

1) $5\frac{1}{2}$

2) 8

3) $4\frac{1}{2}$

4) $4\frac{7}{12}$

5) $6\frac{5}{12}$

6) $8\frac{13}{15}$

7) $6\frac{16}{21}$

8) $7\frac{9}{10}$

9) $11\frac{29}{35}$

10) $14\frac{19}{48}$

11) $1\frac{1}{2}$

12) $2\frac{1}{5}$

13) $1\frac{2}{9}$

14) $1\frac{9}{14}$

15) $3\frac{1}{4}$

16) $3\frac{13}{15}$

17) $3\frac{1}{8}$

18) $2\frac{13}{15}$

19) $3\frac{7}{30}$

20) $4\frac{3}{35}$

21) $5\frac{1}{12}$

22) $2\frac{1}{16}$

23) $4\frac{1}{24}$

24) $2\frac{71}{72}$

25) $5\frac{33}{35}$

26) $6\frac{32}{63}$

Multiplying and Dividing Mixed Numbers

1) $12\frac{3}{8}$

2) $23\frac{1}{9}$

3) $35\frac{15}{16}$

4) $8\frac{2}{3}$

5) 5

6) $6\frac{6}{7}$

7) $30\frac{1}{3}$

8) $7\frac{6}{7}$

9) $21\frac{7}{8}$

10) 11

11) $\frac{4}{7}$

12) $1\frac{1}{4}$

13) $4\frac{2}{9}$

14) $2\frac{23}{24}$

15) 3

16) $\frac{11}{45}$

17) $\frac{19}{84}$

18) $\frac{52}{57}$

19) $4\frac{4}{27}$

20) $2\frac{26}{33}$

21) $3\frac{17}{36}$

22) $\frac{19}{22}$

23) $3\frac{27}{91}$

24) $3\frac{13}{36}$

25) $7\frac{1}{5}$

26) $4\frac{16}{45}$

Chapter 4:
Decimals

Topics that you will practice in this chapter:

- ✓ Adding and Subtracting Decimals
- ✓ Multiplying and Dividing Decimals
- ✓ Comparing Decimals
- ✓ Rounding Decimals

"The study of mathematics, like the Nile, begins in minuteness but ends in magnificence." –
Charles Caleb Colton

Adding and Subtracting Decimals

✎ Add and subtract decimals.

1) 35.19 − 24.28

4) 38.72 − 21.68

7) 86.09 − 35.14

2) 34.29 + 42.58

5) 57.39 + 26.54

8) 54.51 + 32.66

3) 61.20 + 33.75

6) 70.24 − 42.35

9) 114.21 − 88.69

✎ Find the missing number.

10) ___ + 2.8 = 5.4

11) 4.1 + ___ = 5.88

12) 6.45 + ___ = 8

13) 7.25 − ___ = 3.40

14) ___ − 2.35 = 4.25

15) ___ − 19.85 = 6.54

16) 22.15 + ___ = 28.95

17) ___ − 37.16 = 9.42

18) ___ + 24.50 = 34.19

19) 72.40 + ___ = 125.20

GMAS Subject Test Mathematics Grade 6

Multiplying and Dividing Decimals

✎ **Find the product.**

1) $0.5 \times 0.6 =$

2) $3.3 \times 0.4 =$

3) $1.28 \times 0.5 =$

4) $0.35 \times 0.6 =$

5) $1.85 \times 0.6 =$

6) $0.24 \times 0.5 =$

7) $5.25 \times 1.4 =$

8) $18.5 \times 4.6 =$

9) $15.4 \times 6.8 =$

10) $19.5 \times 2.6 =$

11) $32.2 \times 1.5 =$

12) $78.4 \times 4.5 =$

✎ **Find the quotient.**

13) $1.85 \div 10 =$

14) $74.6 \div 100 =$

15) $3.6 \div 3 =$

16) $9.6 \div 0.4 =$

17) $15.5 \div 0.5 =$

18) $32.8 \div 0.2 =$

19) $22.15 \div 1,000 =$

20) $53.55 \div 0.7 =$

21) $322.2 \div 0.2 =$

22) $50.67 \div 0.18 =$

23) $77.4 \div 0.8 =$

24) $27.93 \div 0.03 =$

Comparing Decimals

✏️ **Write the correct comparison symbol (>, < or =).**

1) 0.70 ☐ 0.070

2) 0.049 ☐ 0.49

3) 5.090 ☐ 5.09

4) 2.57 ☐ 2.05

5) 9.03 ☐ 0.930

6) 6.06 ☐ 6.6

7) 7.02 ☐ 7.020

8) 3.04 ☐ 3.2

9) 3.61 ☐ 3.245

10) 0.986 ☐ 0.0986

11) 17.24 ☐ 17.240

12) 0.759 ☐ 0.81

13) 9.040 ☐ 9.40

14) 5.73 ☐ 5.213

15) 9.44 ☐ 9.404

16) 7.17 ☐ 7.170

17) 4.85 ☐ 4.085

18) 9.041 ☐ 9.40

19) 3.033 ☐ 3.030

20) 4.97 ☐ 4.970

GMAS Subject Test Mathematics Grade 6

Rounding Decimals

✎ **Round each decimal to the nearest whole number.**

1) 28.12   3) 16.22   5) 7.95

2) 6.9   4) 8.5   6) 52.7

✎ **Round each decimal to the nearest tenth.**

7) 31.761   9) 94.729   11) 13.219

8) 14.421   10) 77.89   12) 59.89

✎ **Round each decimal to the nearest hundredth.**

13) 8.428   15) 55.3786   17) 62.241

14) 23.812 16) 231.912   18) 19.447

✎ **Round each decimal to the nearest thousandth.**

19) 15.54324   21) 243.8652   23) 67.1983

20) 34.62586   22) 80.4529   24) 72.36788

GMAS Subject Test Mathematics Grade 6

Convert Fraction to Decimal

✎ Write each as a decimal.

1) $\frac{50}{100} =$

2) $\frac{46}{100} =$

3) $\frac{8}{50} =$

4) $\frac{8}{32} =$

5) $\frac{8}{72} =$

6) $\frac{56}{100} =$

7) $\frac{4}{50} =$

8) $\frac{31}{48} =$

9) $\frac{27}{300} =$

10) $\frac{15}{55} =$

11) $\frac{16}{32} =$

12) $\frac{6}{16} =$

13) $\frac{3}{10} =$

14) $\frac{18}{250} =$

15) $\frac{24}{80} =$

16) $\frac{30}{40} =$

17) $\frac{68}{100} =$

18) $\frac{7}{35} =$

19) $\frac{87}{100} =$

20) $\frac{1}{100} =$

21) $\frac{6}{36} =$

22) $\frac{2}{80} =$

WWW.MathNotion.Com

GMAS Subject Test Mathematics Grade 6

Convert Decimal to Percent

✏️ **Write each as a percent.**

1) 0.187 =

2) 0.19 =

3) 2.6 =

4) 0.017 =

5) 0.009 =

6) 0.786 =

7) 0.245 =

8) 0.57 =

9) 0.002 =

10) 0.205 =

11) 0.324 =

12) 84.9 =

13) 3.015 =

14) 0.7 =

15) 2.35 =

16) 0.0367 =

17) 0.0043 =

18) 0.960 =

19) 6.68 =

20) 0.484 =

21) 8.957 =

22) 0.879 =

23) 2.7 =

24) 0.9 =

25) 3.6 =

26) 26.8 =

27) 1.01 =

28) 0.006 =

GMAS Subject Test Mathematics Grade 6

Convert Fraction to Percent

✏️ **Write each as a percent.**

1) $\dfrac{1}{4} =$

2) $\dfrac{3}{8} =$

3) $\dfrac{7}{14} =$

4) $\dfrac{15}{35} =$

5) $\dfrac{12}{28} =$

6) $\dfrac{17}{68} =$

7) $\dfrac{8}{11} =$

8) $\dfrac{14}{30} =$

9) $\dfrac{6}{50} =$

10) $\dfrac{12}{48} =$

11) $\dfrac{5}{34} =$

12) $\dfrac{27}{10} =$

13) $\dfrac{24}{80} =$

14) $\dfrac{16}{25} =$

15) $\dfrac{16}{58} =$

16) $\dfrac{2}{22} =$

17) $\dfrac{32}{88} =$

18) $\dfrac{21}{36} =$

19) $\dfrac{18}{92} =$

20) $\dfrac{6}{60} =$

21) $\dfrac{24}{600} =$

22) $\dfrac{720}{360} =$

GMAS Subject Test Mathematics Grade 6

Answers of Worksheets

Adding and Subtracting Decimals

1) 10.91	6) 27.89	11) 1.78	16) 6.8
2) 76.87	7) 50.95	12) 1.55	17) 46.58
3) 94.95	8) 87.17	13) 3.85	18) 9.69
4) 17.04	9) 25.52	14) 6.6	19) 52.8
5) 83.93	10) 2.6	15) 26.39	

Multiplying and Dividing Decimals

1) 0.3	7) 7.35	13) 0.185	19) 0.02215
2) 1.32	8) 85.1	14) 0.746	20) 76.5
3) 0.64	9) 104.72	15) 1.2	21) 1,611
4) 0.21	10) 50.7	16) 24	22) 281.5
5) 1.11	11) 48.3	17) 31	23) 96.75
6) 0.12	12) 352.8	18) 164	24) 931

Comparing Decimals

1) >	6) <	11) =	16) =
2) <	7) =	12) <	17) >
3) =	8) <	13) <	18) <
4) >	9) >	14) >	19) >
5) >	10) >	15) >	20) =

Rounding Decimals

1) 28	9) 94.7	17) 62.24
2) 7	10) 77.9	18) 19.45
3) 16	11) 13.2	19) 15.543
4) 9	12) 59.9	20) 34.626
5) 8	13) 8.43	21) 243.865
6) 53	14) 23.81	22) 80.453
7) 31.8	15) 55.38	23) 67.198
8) 14.4	16) 231.91	24) 72.368

Convert Fraction to Decimal

1) 0.5	2) 0.46	3) 0.16

GMAS Subject Test Mathematics Grade 6

4) 0.25
5) 0.11
6) 0.56
7) 0.08
8) 0.646
9) 0.09
10) 0.27

11) 0.5
12) 0.375
13) 0.3
14) 0.072
15) 0.3
16) 0.75
17) 0.68

18) 0.2
19) 0.87
20) 0.01
21) 0.166
22) 0.025

Convert Decimal to Percent

1) 18.7%
2) 19%
3) 260%
4) 1.7%
5) 0.9%
6) 78.6%
7) 24.5%
8) 57%
9) 0.2%
10) 20.5%

11) 32.4%
12) 8,490%
13) 301.5%
14) 70%
15) 235%
16) 3.67%
17) 0.43%
18) 96%
19) 668%
20) 48.4%

21) 895.7%
22) 87.9%
23) 270%
24) 90%
25) 360%
26) 2,680%
27) 101%
28) 0.6%

Convert Fraction to Percent

1) 25%
2) 37.5%
3) 50%
4) 42.86%
5) 29.31%
6) 25%
7) 72.72%
8) 46.66%

9) 12%
10) 25%
11) 14.7%
12) 2.7%
13) 30%
14) 64%
15) 27.58%
16) 9.09%

17) 36.36%
18) 58.33%
19) 19.56%
20) 10%
21) 4%
22) 200%

WWW.MathNotion.Com

GMAS Subject Test Mathematics Grade 6

Chapter 5:
Proportions, Ratios, and Percent

Topics that you will practice in this chapter:

- ✓ Simplifying Ratios
- ✓ Proportional Ratios
- ✓ Similarity and Ratios
- ✓ Ratio and Rates Word Problems
- ✓ Percentage Calculations
- ✓ Percent Problems
- ✓ Discount, Tax and Tip

Without mathematics, there's nothing you can do. Everything around you is mathematics. Everything around you is numbers." – Shakuntala Devi

GMAS Subject Test Mathematics Grade 6

Simplifying Ratios

✎ Reduce each ratio.

1) 15: 20 = ___: ___

2) 7: 70 = ___: ___

3) 16: 28 = ___: ___

4) 7: 21 = ___: ___

5) 4: 40 = ___: ___

6) 6: 48 = ___: ___

7) 16: 64 = ___: ___

8) 10: 25 = ___: ___

9) 8: 48 = ___: ___

10) 49: 63 = ___: ___

11) 18: 27 = ___: ___

12) 35: 10 = ___: ___

13) 90: 9 = ___: ___

14) 24: 32 = ___: ___

15) 7: 56 = ___: ___

16) 45: 63 = ___: ___

17) 56: 72 = ___: ___

18) 26: 13 = ___: ___

19) 15: 45 = ___: ___

20) 28: 4 = ___: ___

21) 24: 48 = ___: ___

22) 30: 24 = ___: ___

23) 70: 140 = ___: ___

24) 6: 180 = ___: ___

✎ Write each ratio as a fraction in simplest form.

25) 6: 12 =

26) 30: 50 =

27) 15: 35 =

28) 9: 27 =

29) 8: 24 =

30) 18: 84 =

31) 7: 14 =

32) 7: 35 =

33) 40: 96 =

34) 12: 54 =

35) 44: 52 =

36) 12: 27 =

37) 15: 180 =

38) 39: 143 =

39) 20: 300 =

40) 30: 120 =

41) 56: 42 =

42) 26: 130 =

43) 66: 123 =

44) 70: 630 =

45) 75: 125 =

GMAS Subject Test Mathematics Grade 6

Proportional Ratios

✎ Fill in the blanks; Calculate each proportion.

1) $3:8 = \_\_ : 48$
2) $2:5 = 20: \_\_$
3) $1:9 = \_\_ : 81$
4) $6:7 = 12: \_\_$
5) $9:2 = 63: \_\_$
6) $8:7 = \_\_ : 49$
7) $20:3 = \_\_ : 15$
8) $1:3 = \_\_ : 75$
9) $7:6 = \_\_ : 60$
10) $8:5 = \_\_ : 45$
11) $3:10 = 60: \_\_$
12) $6:11 = 42: \_\_$

✎ State if each pair of ratios form a proportion.

13) $\frac{3}{20}$ and $\frac{9}{60}$
14) $\frac{1}{7}$ and $\frac{6}{42}$
15) $\frac{3}{7}$ and $\frac{24}{56}$
16) $\frac{4}{9}$ and $\frac{12}{18}$
17) $\frac{1}{9}$ and $\frac{12}{81}$
18) $\frac{7}{8}$ and $\frac{21}{28}$
19) $\frac{9}{13}$ and $\frac{27}{39}$
20) $\frac{1}{8}$ and $\frac{8}{64}$
21) $\frac{6}{19}$ and $\frac{30}{85}$
22) $\frac{5}{9}$ and $\frac{40}{81}$
23) $\frac{9}{14}$ and $\frac{108}{168}$
24) $\frac{15}{23}$ and $\frac{360}{552}$

✎ Calculate each proportion.

25) $\frac{20}{25} = \frac{32}{x}, x = \_\_\_$
26) $\frac{1}{8} = \frac{32}{x}, x = \_\_\_$
27) $\frac{15}{5} = \frac{21}{x}, x = \_\_\_$
28) $\frac{1}{7} = \frac{x}{294}, x = \_\_\_$
29) $\frac{7}{9} = \frac{x}{81}, x = \_\_\_$
30) $\frac{1}{5} = \frac{13}{x}, x = \_\_\_$
31) $\frac{9}{5} = \frac{36}{x}, x = \_\_\_$
32) $\frac{6}{13} = \frac{48}{x}, x = \_\_\_$
33) $\frac{5}{8} = \frac{x}{88}, x = \_\_\_$
34) $\frac{4}{15} = \frac{x}{240}, x = \_\_\_$
35) $\frac{9}{19} = \frac{x}{266}, x = \_\_\_$
36) $\frac{7}{15} = \frac{x}{270}, x = \_\_\_$

GMAS Subject Test Mathematics Grade 6

Similarity and Ratios

✎ Each pair of figures is similar. Find the missing side.

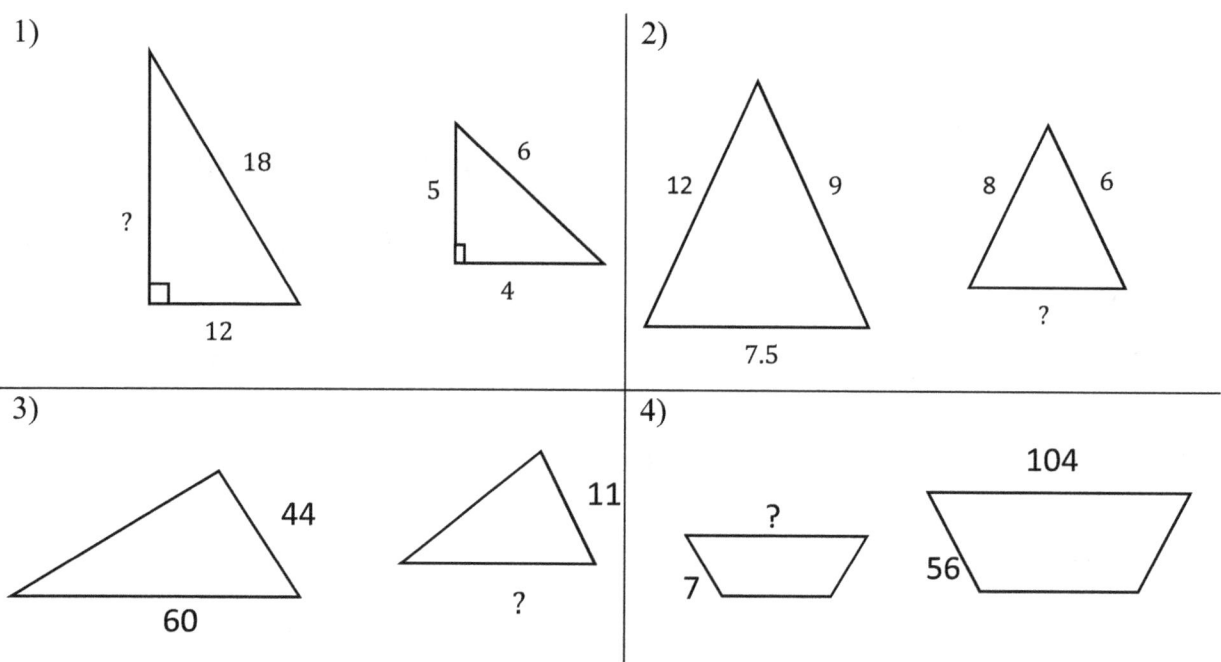

✎ Calculate.

5) Two rectangles are similar. The first is 24 feet wide and 120 feet long. The second is 30 feet wide. What is the length of the second rectangle? _____

6) Two rectangles are similar. One is 5 meters by 36 meters. The longer side of the second rectangle is 90 meters. What is the other side of the second rectangle? _____

7) A building casts a shadow 25 ft long. At the same time a girl 10 ft tall casts a shadow 5 ft long. How tall is the building? _____

8) The scale of a map of Texas is 4 inches: 32 miles. If you measure the distance from Dallas to Martin County as 38.4 inches, approximately how far is Martin County from Dallas? _____

WWW.MathNotion.Com

GMAS Subject Test Mathematics Grade 6

Ratio and Rates Word Problems

✎ Find the answer for each word problem.

1) Mason has 24 red cards and 36 green cards. What is the ratio of Mason's red cards to his green cards? _____

2) In a party, 45 soft drinks are required for every 54 guests. If there are 378 guests, how many soft drinks is required? _____

3) In Mason's class, 42 of the students are tall and 24 are short. In Michael's class 84 students are tall and 48 students are short. Which class has a higher ratio of tall to short students? _____

4) The price of 5 apples at the Quick Market is $4.6. The price of 7 of the same apples at Walmart is $5.95. Which place is the better buy? _____

5) The bakers at a Bakery can make 90 bagels in 3 hours. How many bagels can they bake in 24 hours? What is that rate per hour? _____

6) You can buy 5 cans of green beans at a supermarket for $5.75. How much does it cost to buy 45 cans of green beans? _____

7) The ratio of boys to girls in a class is 4: 7. If there are 32 boys in the class, how many girls are in that class? _____

8) The ratio of red marbles to blue marbles in a bag is 3: 7. If there are 50 marbles in the bag, how many of the marbles are red? _____

GMAS Subject Test Mathematics Grade 6

Percentage Calculations

✎ **Calculate the given percent of each value.**

1) 3% of 60 = ____
2) 20% of 32 = ____
3) 4% of 72 = ____
4) 16% of 32 = ____
5) 25% of 124 = ____
6) 35% of 56 = ____

7) 15% of 20 = ____
8) 14% of 150 = ____
9) 80% of 50 = ____
10) 12% of 115 = ____
11) 72% of 250 = ____
12) 52% of 500 = ____

13) 70% of 400 = ____
14) 27% of 145 = ____
15) 90% of 64 = ____
16) 60% of 55 = ____
17) 22% of 210 = ____
18) 8% of 235 = ____

✎ **Calculate the percent of each given value.**

19) ____% of 25 = 5
20) ____% of 40 = 20
21) ____% of 25 = 2
22) ____% of 50 = 16
23) ____% of 250 = 5

24) ____% of 40 = 32
25) ____% of 125 = 20
26) ____% of 700 = 49
27) ____% of 350 = 49
28) ____% of 500 = 210

✎ **Calculate each percent problem.**

29) A Cinema has 250 seats. 60 seats were sold for the current movie. What percent of seats are empty? ____ %

30) There are 68 boys and 92 girls in a class. 75% of the students in the class take the bus to school. How many students do not take the bus to school? ____

WWW.MathNotion.Com

GMAS Subject Test Mathematics Grade 6

Percent Problems

✎ **Calculate each problem.**

1) 9 is what percent of 45? ___%

2) 60 is what percent of 120? ___%

3) 10 is what percent of 200? ___%

4) 15 is what percent of 125? ___%

5) 10 is what percent of 400? ___%

6) 66 is what percent of 55? ___%

7) 40 is what percent of 160? ___%

8) 40 is what percent of 50? ___%

9) 120 is what percent of 800? ___%

10) 78 is what percent of 120? ___%

11) 36 is what percent of 144? ___%

12) 17 is what percent of 85? ___%

13) 90 is what percent of 900? ___%

14) 36 is what percent of 16? ___%

15) 63 is what percent of 14? ___%

16) 18 is what percent of 60? ___%

17) 126 is what percent of 200? ___%

18) 232 is what percent of 40? ___%

✎ **Calculate each percent word problem.**

19) There are 40 employees in a company. On a certain day, 25 were present. What percent showed up for work? ____%

20) A metal bar weighs 60 ounces. 25% of the bar is gold. How many ounces of gold are in the bar? _____

21) A crew is made up of 12 women; the rest are men. If 15% of the crew are women, how many people are in the crew? _____

22) There are 40 students in a class and 8 of them are girls. What percent are boys? ____%

23) The Royals softball team played 400 games and won 280 of them. What percent of the games did they lose? ____%

WWW.MathNotion.Com

GMAS Subject Test Mathematics Grade 6

Discount, Tax and Tip

🍂 **Find the selling price of each item.**

1) Original price of a computer: $420
Tax: 8% Selling price: $_____

2) Original price of a laptop: $280
Tax: 4% Selling price: $_____

3) Original price of a sofa: $820
Tax: 5% Selling price: $_____

4) Original price of a car: $15,800
Tax: 3.6%   Selling price: $_____

5) Original price of a Table: $250
Tax: 9% Selling price: $_____

6) Original price of a house: $630,000
Tax: 1.8%   Selling price: $_____

7) Original price of a tablet: $450
Discount: 30% Selling price: $____

8) Original price of a chair: $390
Discount: 8%   Selling price: $____

9) Original price of a book: $75
Discount: 42% Selling price: $____

10) Original price of a cellphone: $820
Discount: 23% Selling price: $___

11) Food bill: $45
Tip: 15%   Price: $_____

12) Food bill: $32
Tipp: 20%   Price: $_____

13) Food bill: $90
Tip: 35%   Price: $_____

14) Food bill: $42
Tipp: 12%   Price: $_____

🍂 **Find the answer for each word problem.**

15) Nicolas hired a moving company. The company charged $500 for its services, and Nicolas gives the movers a 40% tip. How much does Nicolas tip the movers? $_____

16) Mason has lunch at a restaurant and the cost of his meal is $90. Mason wants to leave a 25% tip. What is Mason's total bill including tip? $_____

17) The sales tax in Texas is 19.80% and an item costs $350. How much is the tax? $_____

18) The price of a table at Best Buy is $680. If the sales tax is 5%, what is the final price of the table including tax? $_____

WWW.MathNotion.Com

GMAS Subject Test Mathematics Grade 6

Answers of Worksheets

Simplifying Ratios

1) 3 : 4
2) 1 : 10
3) 4 : 7
4) 1 : 3
5) 1 : 10
6) 1 : 8
7) 2 : 8
8) 2 : 5
9) 1 : 6
10) 7 : 9
11) 2 : 3
12) 7 : 2
13) 10 : 1
14) 3 : 4
15) 1 : 8
16) 5 : 7
17) 7 : 9
18) 2 : 1
19) 1 : 3
20) 7 : 1
21) 1 : 2
22) 5 : 4
23) 1 : 2
24) 1 : 30
25) $\frac{1}{2}$
26) $\frac{3}{5}$
27) $\frac{3}{7}$
28) $\frac{1}{3}$
29) $\frac{1}{3}$
30) $\frac{3}{14}$
31) $\frac{1}{2}$
32) $\frac{1}{5}$
33) $\frac{5}{12}$
34) $\frac{2}{9}$
35) $\frac{11}{13}$
36) $\frac{4}{9}$
37) $\frac{1}{12}$
38) $\frac{3}{11}$
39) $\frac{1}{15}$
40) $\frac{1}{4}$
41) $\frac{4}{3}$
42) $\frac{1}{5}$
43) $\frac{22}{41}$
44) $\frac{1}{9}$
45) $\frac{3}{5}$

Proportional Ratios

1) 18
2) 50
3) 9
4) 14
5) 14
6) 56
7) 100
8) 25
9) 70
10) 72
11) 200
12) 77
13) Yes
14) Yes
15) Yes
16) No
17) No
18) No
19) Yes
20) Yes
21) No
22) No
23) Yes
24) Yes
25) 40
26) 256
27) 7
28) 42
29) 63
30) 65
31) 20
32) 104
33) 55
34) 64
35) 126
36) 126

Similarity and ratios

1) 15
2) 5
3) 15
4) 13
5) 150 feet
6) 12.5 meters
7) 50 feet
8) 307.2 miles

Ratio and Rates Word Problems

1) 2 : 3
2) 315

WWW.MathNotion.Com

GMAS Subject Test Mathematics Grade 6

3) The ratio for both classes is 7 to 4.   6) $51.75

4) Walmart is a better buy.   7) 56

5) 720, the rate is 30 per hour.   8) 15

Percentage Calculations

1) 1.8	11) 180	21) 8%
2) 6.4	12) 260	22) 32%
3) 2.88	13) 280	23) 2%
4) 5.12	14) 39.15	24) 80%
5) 31	15) 57.6	25) 16%
6) 19.6	16) 33	26) 7%
7) 3	17) 46.2	27) 14%
8) 21	18) 18.8	28) 42%
9) 40	19) 20%	29) 76%
10) 13.8	20) 50%	30) 40

Percent Problems

1) 20%	9) 15%	17) 63%
2) 50%	10) 65%	18) 580%
3) 5%	11) 25%	19) 62.5%
4) 12%	12) 20%	20) 15 ounces
5) 2.5%	13) 10%	21) 80
6) 120%	14) 225%	22) 80%
7) 25%	15) 450%	23) 30%
8) 80%	16) 30%	

Discount, Tax and Tip

1) $453.60	7) $315.00	13) $121.50
2) $291.20	8) $358.80	14) $47.04
3) $861.00	9) $43.50	15) $200.00
4) $16,368.80	10) $631.40	16) $112.50
5) $272.50	11) $51.75	17) $69.30
6) $641,340	12) $38.40	18) $714.00

WWW.MathNotion.Com

GMAS Subject Test Mathematics Grade 6

Chapter 6:
Exponents and Radicals Expressions

Topics that you will practice in this chapter:

- ✓ Adding and Subtracting Exponents
- ✓ Multiplication Property of Exponents
- ✓ Zero and Negative Exponents
- ✓ Division Property of Exponents
- ✓ Powers of Products and Quotients
- ✓ Negative Exponents and Negative Bases
- ✓ Scientific Notation
- ✓ Square Roots

Mathematics is no more computation than typing is literature.

– John Allen Paulos

GMAS Subject Test Mathematics Grade 6

Adding and Subtracting Exponents

✏️ Solve each problem.

1) $3^2 + 2^5 =$

2) $x^6 + x^6 =$

3) $3b^2 - 2b^2 =$

4) $3 + 4^3 =$

5) $8 - 4^2 =$

6) $4 + 7^1 =$

7) $2x^3 + 3x^3 =$

8) $10^2 + 3^5 =$

9) $4^5 - 2^4 =$

10) $5^2 - 6^0 =$

11) $1^2 - 3^0 =$

12) $7^1 + 2^3 =$

13) $6^1 - 5^3 =$

14) $3^3 + 3^3 =$

15) $9^2 - 8^2 =$

16) $0^{73} + 0^{54} =$

17) $2^2 - 3^2 =$

18) $7^3 - 7^1 =$

19) $8^2 - 6^2 =$

20) $4^2 + 3^2 =$

21) $2^3 + 4^3 =$

22) $10 + 3^3 =$

23) $6x^5 + 8x^5 =$

24) $8^0 + 4^2 =$

25) $3^2 + 3^2 =$

26) $10^2 + 5^2 =$

27) $(\frac{1}{2})^2 + (\frac{1}{2})^2 =$

28) $9^2 + 3^2 =$

GMAS Subject Test Mathematics Grade 6

Multiplication Property of Exponents

✎ Simplify and write the answer in exponential form.

1) $4 \times 4^5 =$

2) $8^4 \times 8 =$

3) $7^3 \times 7^3 =$

4) $9^2 \times 9^2 =$

5) $2^2 \times 2^4 \times 2 =$

6) $5 \times 5^3 \times 5^3 =$

7) $4^3 \times 4^2 \times 4 \times 4 =$

8) $5x \times x =$

9) $x^3 \times x^3 =$

10) $x^7 \times x^2 =$

11) $x^4 \times x^3 \times x^2 =$

12) $10x \times 3x =$

13) $4x^3 \times 4x^3 =$

14) $7x^3 \times x =$

15) $3x^2 \times 4x^2 \times x^2 =$

16) $5x^4 \times x^4 =$

17) $2x^8 \times 2x =$

18) $6x \times x^5 =$

19) $4x^2 \times 6x^6 =$

20) $5yx^3 \times 4x =$

21) $7x^3 \times y^5 x^7 =$

22) $y^2 x^3 \times y^5 x^4 =$

23) $3x^5 \times 4x^3 y^4 =$

24) $4x^4 \times 9x^2 y^5 =$

25) $5x^3 y^4 \times 6x^8 y^2 =$

26) $8x^3 y^6 \times 4xy^3 =$

27) $2xy^5 \times 6x^3 y^3 =$

28) $4x^5 y^2 \times 4x^2 y^8 =$

29) $7x \times 3y^8 x^2 \times y^5 =$

30) $x^3 \times 2y^3 x^4 \times 2y =$

31) $3yx^4 \times 3y^4 x \times 3xy^3 =$

32) $6y^3 \times 2y^2 x^4 \times 10yx^5 =$

WWW.MathNotion.Com

GMAS Subject Test Mathematics Grade 6

Zero and Negative Exponents

✏️ **Evaluate the following expressions.**

1) $1^{-5} =$

2) $4^{-1} =$

3) $0^{10} =$

4) $1^{15} =$

5) $5^{-2} =$

6) $3^{-3} =$

7) $9^{-1} =$

8) $10^{-2} =$

9) $12^{-2} =$

10) $2^{-5} =$

11) $3^{-4} =$

12) $2^{-4} =$

13) $6^{-3} =$

14) $10^{-3} =$

15) $30^{-1} =$

16) $15^{-2} =$

17) $4^{-3} =$

18) $2^{-7} =$

19) $5^{-3} =$

20) $4^{-4} =$

21) $3^{-5} =$

22) $10^{-4} =$

23) $2^{-10} =$

24) $8^{-3} =$

25) $20^{-2} =$

26) $14^{-2} =$

27) $9^{-3} =$

28) $100^{-2} =$

29) $5^{-4} =$

30) $4^{-6} =$

31) $(\frac{1}{4})^{-3} =$

32) $(\frac{1}{6})^{-2} =$

33) $(\frac{1}{7})^{-2} =$

34) $(\frac{2}{3})^{-3} =$

35) $(\frac{1}{13})^{-2} =$

36) $(\frac{7}{12})^{-2} =$

37) $(\frac{1}{6})^{-3} =$

38) $(\frac{1}{300})^{-2} =$

39) $(\frac{2}{9})^{-2} =$

40) $(\frac{7}{5})^{-1} =$

41) $(\frac{13}{23})^{0} =$

42) $(\frac{1}{4})^{-5} =$

WWW.MathNotion.Com

Division Property of Exponents

✎ **Simplify.**

1) $\dfrac{5^6}{5^7} =$

2) $\dfrac{8^8}{8^6} =$

3) $\dfrac{4^5}{4} =$

4) $\dfrac{3}{3^5} =$

5) $\dfrac{x}{x^6} =$

6) $\dfrac{3 \times 3^2}{3^2 \times 3^5} =$

7) $\dfrac{9^4}{9^2} =$

8) $\dfrac{10 \times 10^9}{10^2 \times 10^7} =$

9) $\dfrac{7^5 \times 7^7}{7^4 \times 7^8} =$

10) $\dfrac{15x}{30x^6} =$

11) $\dfrac{3x^9}{4x^4} =$

12) $\dfrac{15x^8}{10x^9} =$

13) $\dfrac{42x^5}{6y^9} =$

14) $\dfrac{36y^8}{4x^4y^5} =$

15) $\dfrac{2x^7}{9x} =$

16) $\dfrac{49x^8y^6}{7x^9} =$

17) $\dfrac{48x^2}{24x^6y^{12}} =$

18) $\dfrac{30yx^5}{6yx^7} =$

19) $\dfrac{19x^7y}{38x^{12}y^4} =$

20) $\dfrac{9x^8}{63x^8} =$

21) $\dfrac{9x^{-9}}{4x^{-3}} =$

GMAS Subject Test Mathematics Grade 6

Powers of Products and Quotients

✍ **Simplify.**

1) $(4^3)^2 =$

2) $(2^3)^4 =$

3) $(2 \times 2^3)^2 =$

4) $(5 \times 5^5)^6 =$

5) $(19^4 \times 19^2)^3 =$

6) $(2^3 \times 2^4)^4 =$

7) $(5 \times 5^2)^2 =$

8) $(4^4)^4 =$

9) $(8x^5)^2 =$

10) $(3x^2 y^4)^4 =$

11) $(7x^5 y^2)^2 =$

12) $(5x^4 y^4)^3 =$

13) $(2x^3 y^3)^5 =$

14) $(10x^3 y^4)^3 =$

15) $(13y^3 y)^2 =$

16) $(5x^6 x^4)^2 =$

17) $(6x^7 y^6)^3 =$

18) $(12x^5 x^7)^2 =$

19) $(2x^4 \times 2x)^4 =$

20) $(2x^4 y^3)^5 =$

21) $(15x^7 y^2)^2 =$

22) $(8x^3 y^5)^3 =$

23) $(3x \times 2y^2)^4 =$

24) $\left(\dfrac{4x}{x^5}\right)^2 =$

25) $\left(\dfrac{x^4 y^5}{x^3 y^5}\right)^9 =$

26) $\left(\dfrac{36xy}{6x^5}\right)^3 =$

27) $\left(\dfrac{x^7}{x^8 y^2}\right)^6 =$

28) $\left(\dfrac{xy^4}{x^3 y^6}\right)^{-3} =$

29) $\left(\dfrac{5xy^8}{x^3}\right)^2 =$

30) $\left(\dfrac{xy^6}{2xy^3}\right)^{-4} =$

WWW.MathNotion.Com

GMAS Subject Test Mathematics Grade 6

Negative Exponents and Negative Bases

✏ Simplify.

1) $-9^{-1} =$

2) $-9^{-2} =$

3) $-2^{-5} =$

4) $-x^{-7} =$

5) $11x^{-1} =$

6) $-8x^{-3} =$

7) $-12x^{-5} =$

8) $-9x^{-8}y^{-6} =$

9) $32x^{-5}y^{-1} =$

10) $10a^{-9}b^{-3} =$

11) $-17x^4y^{-6} =$

12) $-\frac{25}{x^{-5}} =$

13) $-\frac{13x}{a^{-7}} =$

14) $\left(-\frac{1}{3}\right)^{-4} =$

15) $\left(-\frac{3}{4}\right)^{-2} =$

16) $-\frac{14}{a^{-6}b^{-3}} =$

17) $-\frac{7x}{x^{-8}} =$

18) $-\frac{a^{-9}}{b^{-5}} =$

19) $-\frac{11}{x^{-5}} =$

20) $\frac{8b}{-16c^{-6}} =$

21) $\frac{12ab}{a^{-4}b^{-3}} =$

22) $-\frac{8n^{-4}}{32p^{-7}} =$

23) $\frac{16ab^{-6}}{-6c^{-5}} =$

24) $\left(\frac{10a}{5c}\right)^{-4} =$

25) $\left(-\frac{12x}{4yz}\right)^{-3} =$

26) $\frac{8ab^{-7}}{-5c^{-3}} =$

27) $\left(-\frac{x^4}{x^5}\right)^{-5} =$

28) $\left(-\frac{x^{-2}}{7x^3}\right)^{-2} =$

29) $\left(-\frac{x^{-4}}{x^2}\right)^{-6} =$

WWW.MathNotion.Com

Scientific Notation

✎ Write each number in scientific notation.

1) $0.223 =$

2) $0.09 =$

3) $4.5 =$

4) $900 =$

5) $2,000 =$

6) $0.006 =$

7) $33 =$

8) $9,400 =$

9) $1,470 =$

10) $52,000 =$

11) $8,000,000 =$

12) $0.00009 =$

13) $2,158,000 =$

14) $0.0039 =$

15) $0.000075 =$

16) $4,300,000 =$

17) $130,000 =$

18) $4,000,000,000 =$

19) $0.00009 =$

20) $0.0039 =$

✎ Write each number in standard notation.

21) $4 \times 10^{-1} =$

22) $1.2 \times 10^{-3} =$

23) $2.7 \times 10^{5} =$

24) $6 \times 10^{-4} =$

25) $3.6 \times 10^{-3} =$

26) $5.5 \times 10^{5} =$

27) $3.2 \times 10^{4} =$

28) $3.88 \times 10^{6} =$

29) $7 \times 10^{-6} =$

30) $4.2 \times 10^{-7} =$

GMAS Subject Test Mathematics Grade 6

Square Roots

✎ **Find the value each square root.**

1) $\sqrt{16} =$ ___

2) $\sqrt{25} =$ ___

3) $\sqrt{1} =$ ___

4) $\sqrt{64} =$ ___

5) $\sqrt{0} =$ ___

6) $\sqrt{196} =$ ___

7) $\sqrt{4} =$ ___

8) $\sqrt{256} =$ ___

9) $\sqrt{36} =$ ___

10) $\sqrt{289} =$ ___

11) $\sqrt{169} =$ ___

12) $\sqrt{144} =$ ___

13) $\sqrt{100} =$ ___

14) $\sqrt{1,600} =$ ___

15) $\sqrt{2,500} =$ ___

16) $\sqrt{324} =$ ___

17) $\sqrt{529} =$ ___

18) $\sqrt{20} =$ ___

19) $\sqrt{625} =$ ___

20) $\sqrt{18} =$ ___

21) $\sqrt{50} =$ ___

22) $\sqrt{1,024} =$ ___

23) $\sqrt{160} =$ ___

24) $\sqrt{32} =$ ___

✎ **Evaluate.**

25) $\sqrt{4} \times \sqrt{25} =$ _____

26) $\sqrt{36} \times \sqrt{49} =$ _____

27) $\sqrt{6} \times \sqrt{6} =$ _____

28) $\sqrt{13} \times \sqrt{13} =$ _____

29) $2\sqrt{5} \times 3\sqrt{5} =$ _____

30) $\sqrt{12} \times \sqrt{3} =$ _____

31) $\sqrt{13} + \sqrt{13} =$ _____

32) $\sqrt{10} + 2\sqrt{10} =$ _____

33) $12\sqrt{7} - 10\sqrt{7} =$ _____

34) $4\sqrt{10} \times 2\sqrt{10} =$ _____

35) $5\sqrt{3} \times 8\sqrt{3} =$ _____

36) $6\sqrt{3} - \sqrt{12} =$ _____

WWW.MathNotion.Com

GMAS Subject Test Mathematics Grade 6

Answers of Worksheets

Add and Subtract Exponents.

1) 41
2) $2x^6$
3) b^2
4) 67
5) −8
6) 11
7) $5x^3$
8) 343
9) 1,008
10) 24
11) 0
12) 15
13) −119
14) 54
15) 17
16) 0
17) −5
18) 336
19) 28
20) 25
21) 72
22) 37
23) $14x^5$
24) 17
25) 18
26) 125
27) $\frac{1}{2}$
28) 90

Multiplication Property of Exponents

1) 4^6
2) 8^5
3) 7^6
4) 9^4
5) 2^7
6) 5^7
7) 4^7
8) $5x^2$
9) x^6
10) x^9
11) x^9
12) $30x^2$
13) $16x^6$
14) $7x^4$
15) $12x^6$
16) $5x^8$
17) $4x^9$
18) $6x^6$
19) $24x^8$
20) $20x^4y$
21) $7x^{10}y^5$
22) x^7y^7
23) $12x^8y^4$
24) $36x^6y^5$
25) $30x^{11}y^6$
26) $32x^4y^9$
27) $12x^4y^8$
28) $16x^7y^{10}$
29) $21x^3y^{13}$
30) $4x^7y^4$
31) $27x^6y^8$
32) $120x^9y^6$

Zero and Negative Exponents

1) 1
2) $\frac{1}{4}$
3) 0
4) 1
5) $\frac{1}{25}$
6) $\frac{1}{27}$
7) $\frac{1}{9}$
8) $\frac{1}{100}$
9) $\frac{1}{144}$
10) $\frac{1}{32}$
11) $\frac{1}{81}$
12) $\frac{1}{16}$
13) $\frac{1}{216}$
14) $\frac{1}{1,000}$
15) $\frac{1}{30}$
16) $\frac{1}{225}$
17) $\frac{1}{64}$
18) $\frac{1}{128}$
19) $\frac{1}{125}$
20) $\frac{1}{256}$
21) $\frac{1}{243}$
22) $\frac{1}{10,000}$
23) $\frac{1}{1,024}$
24) $\frac{1}{512}$
25) $\frac{1}{400}$

WWW.MathNotion.Com

GMAS Subject Test Mathematics Grade 6

26) $\frac{1}{196}$
27) $\frac{1}{729}$
28) $\frac{1}{10,000}$
29) $\frac{1}{625}$

30) $\frac{1}{4,096}$
31) 64
32) 36
33) 49
34) $\frac{27}{8}$

35) 169
36) $\frac{144}{49}$
37) 216
38) 90,000
39) $\frac{81}{4}$

40) $\frac{5}{7}$
41) 1
42) 1,024

Division Property of Exponents

1) $\frac{1}{5}$
2) 8^2
3) 4^4
4) $\frac{1}{3^4}$
5) $\frac{1}{x^5}$
6) $\frac{1}{3^4}$

7) 9^2
8) 10
9) 1
10) $\frac{1}{2x^5}$
11) $\frac{3x^5}{4}$
12) $\frac{3}{2x}$

13) $\frac{7x^5}{y^9}$
14) $\frac{9y^3}{x^4}$
15) $\frac{2x^6}{9}$
16) $\frac{7y^6}{x}$
17) $\frac{2}{x^4 y^{12}}$

18) $\frac{5}{x^2}$
19) $\frac{1}{2x^5 y^3}$
20) $\frac{1}{7}$
21) $\frac{9}{4x^6}$

Powers of Products and Quotients

1) 4^6
2) 2^{12}
3) 2^8
4) 5^{36}
5) 19^{18}
6) 2^{28}
7) 5^6
8) 4^{16}
9) $64x^{10}$
10) $81x^8 y^{16}$
11) $49x^{10} y^4$

12) $125x^{12} y^{12}$
13) $32x^{15} y^{15}$
14) $1,000x^9 y^{12}$
15) $169y^8$
16) $25x^{20}$
17) $216x^{21} y^{18}$
18) $144x^{24}$
19) $256x^{20}$
20) $32x^{20} y^{15}$
21) $225x^{14} y^4$
22) $512x^9 y^{15}$

23) $1,296x^4 y^8$
24) $\frac{16}{x^8}$
25) x^9
26) $\frac{216y^3}{x^{12}}$
27) $\frac{1}{x^6 y^{12}}$
28) $x^6 y^6$
29) $\frac{25y^{16}}{x^4}$
30) $\frac{16}{y^{12}}$

Negative Exponents and Negative Bases

1) $-\frac{1}{9}$
2) $-\frac{1}{81}$
3) $-\frac{1}{32}$

4) $-\frac{1}{x^7}$
5) $\frac{11}{x}$
6) $-\frac{8}{x^3}$

7) $-\frac{12}{x^5}$
8) $-\frac{9}{x^8 y^6}$
9) $\frac{32}{x^5 y}$

GMAS Subject Test Mathematics Grade 6

10) $\frac{10}{a^9 b^3}$

11) $-\frac{17x^4}{y^6}$

12) $-25x^5$

13) $-13xa^7$

14) 81

15) $\frac{16}{9}$

16) $-14a^6 b^3$

17) $-7x^9$

18) $-\frac{b^5}{a^9}$

19) $-11x^5$

20) $-\frac{bc^6}{2}$

21) $12a^5 b^4$

22) $-\frac{p^7}{4n^4}$

23) $-\frac{8ac^5}{3b^6}$

24) $\frac{c^4}{16a^4}$

25) $\frac{y^3 z^3}{27x^3}$

26) $-\frac{8ac^3}{5b^7}$

27) $-x^5$

28) $49x^{10}$

29) x^{36}

Scientific Notation

1) 2.23×10^{-1}
2) 9×10^{-2}
3) 4.5×10^0
4) 9×10^2
5) 2×10^3
6) 6×10^{-3}
7) 3.3×10^1
8) 9.4×10^3
9) 1.47×10^3
10) 5.2×10^4

11) 8×10^6
12) 9×10^{-5}
13) 2.158×10^6
14) 3.9×10^{-3}
15) 7.5×10^{-5}
16) 4.3×10^6
17) 1.3×10^5
18) 4×10^9
19) 9×10^{-5}
20) 3.9×10^{-3}

21) 0.4
22) 0.0012
23) $270,000$
24) 0.0006
25) 0.0036
26) $550,000$
27) $32,000$
28) $3,880,000$
29) 0.000007
30) 0.00000042

Square Roots

1) 4
2) 5
3) 1
4) 8
5) 0
6) 14
7) 2
8) 16
9) 6

10) 17
11) 13
12) 12
13) 10
14) 40
15) 50
16) 18
17) 23
18) $2\sqrt{5}$

19) 25
20) $3\sqrt{2}$
21) $5\sqrt{2}$
22) 32
23) $4\sqrt{10}$
24) $4\sqrt{2}$
25) 10
26) 42
27) 6

28) 13
29) 30
30) 6
31) $2\sqrt{13}$
32) $3\sqrt{10}$
33) $2\sqrt{7}$
34) 80
35) 120
36) $4\sqrt{3}$

GMAS Subject Test Mathematics Grade 6

Chapter 7 : Measurements

Topics that you will learn in this chapter:

- ✓ Reference Measurement
- ✓ Metric Length
- ✓ Customary Length
- ✓ Metric Capacity
- ✓ Customary Capacity
- ✓ Metric Weight and Mass
- ✓ Customary Weight and Mass
- ✓ Temperature
- ✓ Time

"It's not that I'm so smart, it's just that I stay with problems longer." -Albert Einstein

GMAS Subject Test Mathematics Grade 6

Reference Measurement

LENGTH	
Customary	**Metric**
1 mile (mi) = 1,760 yards (yd)	1 kilometer (km) = 1,000 meters (m)
1 yard (yd) = 3 feet (ft)	1 meter (m) = 100 centimeters (cm)
1 foot (ft) = 12 inches (in.)	1 centimeter(cm) = 10 millimeters(mm)

VOLUME AND CAPACITY	
Customary	**Metric**
1 gallon (gal) = 4 quarts (qt)	1 liter (L) = 1,000 milliliters (mL)
1 quart (qt) = 2 pints (pt.)	
1 pint (pt.) = 2 cups (c)	
1 cup (c) = 8 fluid ounces (Fl oz)	

WEIGHT AND MASS	
Customary	**Metric**
1 ton (T) = 2,000 pounds (lb.)	1 kilogram (kg) = 1,000 grams (g)
1 pound (lb.) = 16 ounces (oz)	1 gram (g) = 1,000 milligrams (mg)

Time
1 year = 12 months
1 year = 52 weeks
1 week = 7 days
1 day = 24 hours
1 hour = 60 minutes
1 minute = 60 seconds

WWW.MathNotion.Com

GMAS Subject Test Mathematics Grade 6

Metric Length Measurement

✎ Convert to the units.

1) 3×10^3 mm = _____ cm

2) 0.95 m = _____ mm

3) 0.08 m = _____ cm

4) 2.25 km = _____ m

5) 7,800 mm = _____ m

6) 9,100 cm = _____ m

7) 5.83 m = _____ cm

8) 2×10^5 mm = _____ cm

9) 8×10^3 mm = _____ m

10) 0.003 km = _____ mm

11) 0.7 km = _____ m

12) 0.011 m = _____ cm

13) 125×10^5 m = _____ km

14) 78×10^4 m = _____ km

Customary Length Measurement

✎ Convert to the units.

1) 15 ft = _____ in

2) 1.5 ft = _____ in

3) 4.8 yd = _____ ft

4) 0.82 yd = _____ ft

5) 17×10^{-3} yd = _____ in

6) 0.5 mi = _____ in

7) 1,746 in = _____ yd

8) 3.24 in = _____ yd

9) 3,960 yd = _____ mi

10) 42.55 yd = _____ in

11) 5×10^{-2} mi = _____ yd

12) 87,120 ft = _____ mi

13) 2.52 in = _____ ft

14) 29.3 yd = _____ feet

15) 0.612 in = _____ ft

16) 1.3 mi = _____ ft

WWW.MathNotion.Com

GMAS Subject Test Mathematics Grade 6

Metric Capacity Measurement

✎ Convert the following measurements.

1) 1.58 l = _____ ml
2) 0.504 l = _____ ml
3) 3.04 l = _____ ml
4) 0.005 l = _____ ml
5) 121.56 l = _____ ml
6) 0.0459 l = _____ ml
7) 4.2×10^5 ml = _____ l
8) 3.12×10^3 ml = _____ l
9) $1,889 \times 10^2$ ml = _____ l
10) 250 ml = _____ l
11) 656,160 ml = _____ l
12) 0.54×10^4 ml = _____ l

Customary Capacity Measurement

✎ Convert the following measurements.

1) 0.7 gal = _____ qt.
2) 3.2 gal = _____ pt.
3) 0.75 gal = _____ c.
4) 15.5 pt. = _____ c
5) 18.2 c = _____ fl oz
6) 9.02 qt = _____ pt.
7) 1.05 qt = _____ c
8) 158 pt. = _____ c
9) 9.6×10^3 c = _____ gal
10) 203.2 pt. = _____ gal
11) 12.4 qt = _____ gal
12) 115.6 pt. = _____ qt
13) 4,880 c = _____ qt
14) 113.6 c = _____ pt.
15) 0.036 qt = _____ gal
16) 522.4 pt. = _____ qt
17) 5.8 gal = _____ pt.
18) 0.002 qt = _____ c
19) 672 c = _____ gal
20) 72.96 fl oz = _____ c

WWW.MathNotion.Com

GMAS Subject Test Mathematics Grade 6

Metric Weight and Mass Measurement

✏️ **Convert.**

1) 0.712 kg = _____ g

2) 54.01 kg = _____ g

3) 9.8×10^{-5} kg = _____ g

4) 0.012 kg = _____ g

5) 120.02 kg = _____ g

6) 1.199 kg = _____ g

7) 0.0055 kg = _____ g

8) 9×10^4 g = _____ kg

9) 3.5×10^5 g = _____ kg

10) 0.008×10^4 g = _____ kg

11) 15,010 g = _____ kg

12) 12.1×10^4 g = _____ kg

13) 4,155,200 g = _____ kg

14) 402×10^2 g = _____ kg

Customary Weight and Mass Measurement

✏️ **Convert.**

1) 36×10^2 lb. = _____ T

2) 0.022×10^4 lb. = _____ T

3) 215,000 lb. = _____ T

4) 12,600 lb. = _____ T

5) 0.015 lb. = _____ oz

6) 1.6 lb. = _____ oz

7) 0.021 lb. = _____ oz

8) 5.2 T = _____ lb.

9) 6.8×10^{-3} T = _____ lb.

10) 156×10^{-2} T = _____ lb.

11) 0.017 T = _____ lb.

12) 1.085 T = _____ oz

13) 0.006 T = _____ oz

14) 209.92 oz = _____ lb.

GMAS Subject Test Mathematics Grade 6

Temperature

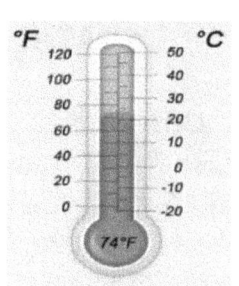

✏ **Convert Fahrenheit into Celsius.**

1) 35.6°F = ___ °C

2) 54.5°F = ___ °C

3) −25.6°F = ___ °C

4) 62.6°F = ___ °C

5) 120.2°F = ___ °C

6) 174.2°F = ___ °C

7) 51.8°F = ___ °C

8) 193.1°F = ___ °C

9) 221°F = ___ °C

10) 60.44°F = ___ °C

11) 48.2°F = ___ °C

12) 134.6°F = ___ °C

✏ **Convert Celsius into Fahrenheit.**

13) 18.2°C = ___ °F

14) 88.8°C = ___ °F

15) 250°C = ___ °F

16) 52°C = ___ °F

17) 10°C = ___ °F

18) −24°C = ___ °F

19) 6°C = ___ °F

20) 15.6°C = ___ °F

21) 33°C = ___ °F

22) 61°C = ___ °F

23) 113.5°C = ___ °F

24) 28°C = ___ °F

WWW.MathNotion.Com

GMAS Subject Test Mathematics Grade 6

Time

🔖 **Convert to the units.**

1) 18.5 hr. = _____ min

2) 26 year = _____ week

3) 0.2 hr. = _____ sec

4) 12.5 min = _____ sec

5) 1.2×10^5 min = _____ hr

6) 1,460 day = _____ year

7) 3 year = _____ hr.

8) 51 day = _____ hr

9) 5 day = _____ min

10) 552 min = _____ hr

11) 32.25 year = _____ month

12) 6,480 sec = _____ min

13) 264 hr = _____ day

14) 22 weeks = _____ day

🔖 **How much time has passed?**

1) From 1:45 A.M. to 4:55 A.M.: ____ hours and ____ minutes.

2) From 1:25 A.M. to 6:05 A.M.: ____ hours and ____ minutes.

3) It's 7:15 P.M. What time was 4 hours ago? _____ O'clock

4) 3:05 A.M to 6:55 AM: _____ hours and _____ minutes.

5) 3:45 A.M to 5:15 AM: _____ hours and _____ minutes.

6) 8:05 A.M. to 11:20 AM. = _____ hour(s) and _____ minutes.

7) 10:55 A.M. to 1:25 PM. = _____ hour(s) and _____ minutes

8) 6:18 A.M. to 6:52 A.M. = _____ minutes

9) 3:54 A.M. to 4:08 A.M. = _____ minutes

WWW.MathNotion.Com

GMAS Subject Test Mathematics Grade 6

Answers of Worksheets

Metric length

1) 300 cm
2) 950 mm
3) 8 cm
4) 2,250 m
5) 7.8 m
6) 91 m
7) 583 cm
8) 20,000 cm
9) 8 m
10) 3,000 mm
11) 700 m
12) 1.1 cm
13) 12,500 km
14) 780 km

Customary Length

1) 180
2) 18
3) 14.4
4) 2.46
5) 0.612
6) 31,680
7) 48.5
8) 0.09
9) 2.25
10) 1,531.8
11) 88
12) 16.5
13) 0.21
14) 87.9
15) 0.051
16) 6,864

Metric Capacity

1) 1,580 ml
2) 504 ml
3) 3,040 ml
4) 5 ml
5) 121,560 ml
6) 45.9 ml
7) 420 L
8) 3.12 L
9) 188.9 L
10) 0.25 L
11) 656.16 L
12) 5.4 L

Customary Capacity

1) 2.8 qt
2) 25.6 pt.
3) 12 c
4) 31 c
5) 145.6 fl oz
6) 18.04 pt.
7) 4.2 c
8) 316 c
9) 600 gal
10) 25.4 gal
11) 3.1 gal
12) 57.8 qt
13) 1,220 qt
14) 56.8 pt.
15) 0.009 gal
16) 261.2 qt
17) 46.4 pt.
18) 0.008 c
19) 42 gal
20) 9.12 c

Metric Weight and Mass

1) 712 g
2) 54,010 g
3) 0.098 g
4) 12 g
5) 120,020 g
6) 1,199 g
7) 5.5 g
8) 90 kg
9) 350 kg
10) 0.08 kg
11) 15.01 kg
12) 121 kg
13) 4,155.2 kg
14) 40.2 kg

GMAS Subject Test Mathematics Grade 6

Customary Weight and Mass

1) 1.8 T	6) 25.6 oz	11) 34 lb.
2) 0.11 T	7) 0.336 oz	12) 34,720 oz
3) 107.5 T	8) 10,400 lb.	13) 192 oz
4) 6.3 T	9) 13.6 lb.	14) 13.12 lb
5) 0.24 oz	10) 3,120 lb.	

Temperature

1) 2°C	9) 105°C	17) 50°F
2) 12.5°C	10) 15.8°C	18) −11.2°F
3) −32°C	11) 9°C	19) 42.8°F
4) 17°C	12) 57°C	20) 60.08°F
5) 49°C	13) 64.76°F	21) 91.4°F
6) 79°C	14) 191.84°F	22) 141.8°F
7) 11°C	15) 482°F	23) 236.3°F
8) 89.5°C	16) 125.6°F	24) 82.4°F

Time - Convert.

1) 1,110 min	6) 4 year	11) 387 months
2) 1,352 weeks	7) 26,280 hr	12) 108 min
3) 720 sec	8) 1,224 hr	13) 11 days
4) 750 sec	9) 7,200 min	14) 154 days
5) 2,000 hr	10) 9.2 hr	

Time - Gap

1) 3:10	4) 3:50	7) 2:30
2) 4:40	5) 1:30	8) 34 minutes
3) 3:15 P.M.	6) 3:15	9) 14 minutes

GMAS Subject Test Mathematics Grade 6

Chapter 8:
Algebraic Expressions

Topics that you will practice in this chapter:

- ✓ Find a rule!
- ✓ Translate Phrases into an Algebraic Statement
- ✓ Simplifying Variable Expressions
- ✓ The Distributive Property
- ✓ Evaluating One Variable Expressions
- ✓ Evaluating Two Variables Expressions
- ✓ Combining like Terms

Mathematics is, as it were, a sensuous logic, and relates to philosophy as do the arts, music, and plastic art to poetry. — *K. Shegel*

Find a Rule!

✏ **Complete the output.**

1- **Rule:** the output is $x - 10.5$

Input	x	15	18	27	32.25	48.5
Output	y					

1) **Rule:** the output is $x \times 5\frac{1}{3}$

Input	x	3	9	15	21	33
Output	y					

2- **Rule:** the output is $x \div 9$

Input	x	513	387	342	198	126
Output	y					

✏ **Find a rule to write an expression.**

3- **Rule:** _____

Input	x	4	14	19	24
Output	y	10	35	47.5	60

4- **Rule:** _____

Input	x	5	13	19.6	34.5
Output	y	14.4	22.4	29	43.9

5- **Rule:** _____

Input	x	72	96	132	230.4
Output	y	9	12	16.5	28.8

GMAS Subject Test Mathematics Grade 6

Translate Phrases into an Algebraic Statement

✏️ Write an algebraic expression for each phrase.

1) 9 multiplied by x. _____

2) Subtract 11 from y. _____

3) 19 divided by x. _____

4) 38 decreased by y. _____

5) Add y to 40. _____

6) The square of 6. _____

7) x raised to the fifth power. _____

8) The sum of six and a number. _____

9) The difference between fifty–seven and y. _____

10) The quotient of nine and a number. _____

11) The quotient of the square of x and 25. _____

12) The difference between x and 6 is 19. _____

13) 10 times a reduced by the square of b. _____

14) Subtract the product of a and b from 41. _____

Simplifying Variable Expressions

✏️ **Simplify each expression.**

1) $3(x + 5) =$

2) $(-4)(7x - 5) =$

3) $11x + 5 - 6x =$

4) $-4 - 2x^2 - 6x^2 =$

5) $7 + 13x^2 + 3 =$

6) $3x^2 + 7x + 15x^2 =$

7) $3x^2 - 12x^2 + 4x =$

8) $4x^2 - 8x - 2x =$

9) $6x + 7(3 - 4x) =$

10) $8x + 4(15x - 3) =$

11) $6(-3x - 9) - 17 =$

12) $-11x^2 - (-5x) =$

13) $2x + 7 + 5 - 8x =$

14) $7 + 6x - 11 - 5x =$

15) $27x + 8 - 13 - 5x =$

16) $(-11)(-5x + 2) - 41x =$

17) $19x - 4(4 - 2x) =$

18) $16x + 3(3x + 6) + 10 =$

19) $5(-2x - 4) - 13x =$

20) $16x - 3x(x + 10) =$

21) $17x + 5x(2 - 4x) =$

22) $5x(-4x - 7) + 20x =$

23) $25x - 19 + 4x^2 =$

24) $6x(x - 11) + 25 =$

25) $4x - 5 + 15x + 3x^2 =$

26) $-7x^2 - 11x - 9x =$

27) $10x - 9x^2 - 3x^2 - 7 =$

28) $13 + 3x^2 - 9x^2 - 21x =$

29) $22x + 10x^2 - 15x + 17 =$

30) $4x^2 + 25x + 21x^2 =$

31) $29 - 12x^2 - 23x - 4x^2 =$

32) $22x - 19x - 9x^2 + 30 =$

The Distributive Property

✏ Use the distributive property to simply each expression.

1) $4(1 + 2x) =$

2) $2(4 + 7x) =$

3) $3(4x - 4) =$

4) $(2x - 5)(-6) =$

5) $(-3)(x + 6) =$

6) $(4 + 3x)2 =$

7) $(-5)(8 - 3x) =$

8) $-(-5 - 7x) =$

9) $(-6x + 3)(-3) =$

10) $(-4)(x - 7) =$

11) $-(5 - 3x) =$

12) $3(9 + 4x) =$

13) $6(4 + 3x) =$

14) $(-5x + 3)2 =$

15) $(5 - 8x)(-3) =$

16) $(-12)(3x + 3) =$

17) $(5 - 3x)6 =$

18) $4(2 + 6x) =$

19) $8(7x - 3) =$

20) $(-2x + 3)4 =$

21) $(7 - 5x)(-9) =$

22) $(-10)(x - 8) =$

23) $(11 - 4x)3 =$

24) $(-6)(10x - 4) =$

25) $(3 - 9x)(-7) =$

26) $(-9)(x + 9) =$

27) $(-3 + 5x)(-7) =$

28) $(-5)(8 - 10x) =$

29) $12(4x - 8) =$

30) $(-10x + 13)(-3) =$

31) $(-8)(3x - 2) + 4(x + 5) =$

32) $(-8)(x + 4) - (6 + 5x) =$

GMAS Subject Test Mathematics Grade 6

Evaluating One Variable Expressions

✎ **Evaluate each expression using the value given.**

1) $8 - x, x = 5$

2) $x - 9, x = 5$

3) $5x + 4, x = 3$

4) $x - 13, x = -4$

5) $12 - x, x = 4$

6) $x + 2, x = 6$

7) $4x + 8, x = 3$

8) $x + (-7), x = -8$

9) $4x + 5, x = 2$

10) $3x + 9, x = -2$

11) $15 + 3x - 7, x = 2$

12) $17 - 3x, x = 3$

13) $8x - 9, x = 4$

14) $5x + 4, x = -3$

15) $10x + 5, x = 3$

16) $14 - 4x, x = -6$

17) $3(5x + 3), x = 9$

18) $4(-3x - 6), x = 3$

19) $7x - 2x + 12, x = 4$

20) $(5x + 6) \div 2, x = 8$

21) $(x + 18) \div 10, x = 12$

22) $5x - 12 + 3x, x = -3$

23) $(6 - 4x)(-3), x = -4$

24) $9x^2 + 3x - 6, x = 2$

25) $x^2 - 10x, x = -5$

26) $3x(7 - 2x), x = 2$

27) $12x + 6 - 2x^2, x = -4$

28) $(-3)(4x - 8 + 3x), x = 3$

29) $(-6) + \frac{x}{4} + 3x, x = 16$

30) $(-6) + \frac{x}{5}, x = 35$

31) $\left(-\frac{45}{x}\right) - 7 + 2x, x = 9$

32) $\left(-\frac{21}{x}\right) - 12 + 4x, x = 7$

WWW.MathNotion.Com

GMAS Subject Test Mathematics Grade 6

Combining like Terms

✏️ **Simplify each expression.**

1) $11x + 3x + 6 =$

2) $8(2x - 6) =$

3) $18x - 7x + 11 =$

4) $(-4)(6x - 7) =$

5) $22x - 10x - 5 =$

6) $32x - 13 + 8x =$

7) $15 - (8x - 11) =$

8) $-24x + 17 - 11x =$

9) $12x - 8 - 6x + 9 =$

10) $21x + 5 - 36 + 12x =$

11) $28x + 3x - 11 =$

12) $(-3x + 4)5 =$

13) $2 + 4x + 9x - 8 =$

14) $6(2x - 5x) - 4 =$

15) $4(5x + 11) + 3x =$

16) $x - 14 - 11x =$

17) $5(10 + 9x) - 8x =$

18) $42x + 17 - 23x =$

19) $(-7x) + 19 + 20x =$

20) $(-7x) - 33 + 29x =$

21) $4(5x + 3) - 19x =$

22) $5(6 - 2x) - 15x =$

23) $-24x + (11 - 18x) =$

24) $(-9) - (6)(7x + 3) =$

25) $(-1)(8x - 10) - 21x =$

26) $-36x + 14 + 27x - 5x =$

27) $3(-13x + 6) - 17x =$

28) $-5x - 42 + 32x =$

29) $37x - 19x + 15 - 9x =$

30) $3(5x + 7x) - 31 =$

31) $14 - 6x - 15 - 9x =$

32) $-2(-5x - 7x) + 27x =$

GMAS Subject Test Mathematics Grade 6

Answers of Worksheets

Find a rule.

1)
Input	x	15	18	27	32.25	48.5
Output	y	4.5	7.5	16.5	21.75	38

2)
Input	x	3	9	15	21	33
Output	y	16	48	80	112	176

3)
Input	x	513	387	342	198	126
Output	y	57	43	38	22	14

4) $y = 2.5x$ 5) $y = x + 9.4$ 6) $y = x \div 8$

Translate Phrases into an Algebraic Statement

1) $9x$
2) $y - 11$
3) $\frac{19}{x}$
4) $38 - y$
5) $y + 40$
6) 6^2
7) x^5
8) $6 + x$
9) $57 - y$
10) $\frac{9}{x}$
11) $\frac{x^2}{25}$
12) $x - 6 = 19$
13) $10a - b^2$
14) $41 - ab$

Simplifying Variable Expressions

1) $3x + 15$
2) $-28x + 20$
3) $5x + 5$
4) $-8x^2 - 4$
5) $13x^2 + 10$
6) $18x^2 + 7x$
7) $-9x^2 + 4x$
8) $4x^2 - 10x$
9) $-22x + 21$
10) $68x - 12$
11) $-18x - 71$
12) $-11x^2 + 5x$
13) $-6x + 12$
14) $x - 4$
15) $22x - 5$
16) $14x - 22$
17) $27x - 16$
18) $25x + 28$
19) $-23x - 20$
20) $-3x^2 - 14x$
21) $-20x^2 + 27x$
22) $-20x^2 - 15x$
23) $4x^2 + 25x - 19$
24) $6x^2 - 66x + 25$
25) $3x^2 + 19x - 5$
26) $-7x^2 - 20x$
27) $-12x^2 + 10x - 7$
28) $-6x^2 - 21x + 13$
29) $10x^2 + 7x + 17$
30) $25x^2 + 25x$
31) $-16x^2 - 23x + 29$
32) $-9x^2 + 3x + 30$

The Distributive Property

1) $8x + 4$
2) $14x + 8$
3) $12x - 12$
4) $-12x + 30$
5) $-3x - 18$
6) $6x + 8$
7) $15x - 40$
8) $7x + 5$

WWW.MathNotion.Com

GMAS Subject Test Mathematics Grade 6

9) $18x - 9$
10) $-4x + 28$
11) $3x - 5$
12) $12x + 27$
13) $18x + 24$
14) $-10x + 6$

15) $24x - 15$
16) $-36x - 36$
17) $-18x + 30$
18) $24x + 8$
19) $56x - 24$
20) $-8x + 12$

21) $45x - 63$
22) $-10x + 80$
23) $-12x + 33$
24) $-60x + 24$
25) $63x - 21$
26) $-9x - 81$

27) $-35x + 21$
28) $50x - 40$
29) $48x - 96$
30) $30x - 39$
31) $-20x + 36$
32) $-13x - 38$

Evaluating One Variables

1) 3
2) -4
3) 19
4) -17
5) 8
6) 8
7) 20
8) -15

9) 13
10) 3
11) 14
12) 8
13) 23
14) -11
15) 35
16) 38

17) 144
18) -60
19) 32
20) 23
21) 3
22) -36
23) -66
24) 36

25) 75
26) 18
27) -74
28) -39
29) 46
30) 1
31) 6
32) 13

Combining like Terms

1) $14x + 6$
2) $16x - 48$
3) $11x + 11$
4) $-24x + 28$
5) $12x - 5$
6) $40x - 13$
7) $-8x + 26$
8) $-35x + 17$

9) $6x + 1$
10) $33x - 31$
11) $31x - 11$
12) $-15x + 20$
13) $13x - 6$
14) $-18x - 4$
15) $23x + 44$
16) $-10x - 14$

17) $37x + 50$
18) $19x + 17$
19) $13x + 19$
20) $22x - 33$
21) $x + 12$
22) $-25x + 30$
23) $-42x + 11$
24) $-42x - 27$

25) $-29x + 10$
26) $-14x + 14$
27) $-56x + 18$
28) $27x - 42$
29) $9x + 15$
30) $36x - 31$
31) $-15x - 1$
32) $51x$

GMAS Subject Test Mathematics Grade 6

Chapter 9 :
Equations and Inequalities

Topics that you will practice in this chapter:

- ✓ One-Step Equations
- ✓ Two-Step Equations
- ✓ Multi-Step Equations
- ✓ Graphing Single-Variable Inequalities
- ✓ One-Step Inequalities
- ✓ Two-Step Inequalities
- ✓ Multi-Step Inequalities

"Life is a math equation. In order to gain the most, you have to know how to convert negatives into positives." - Anonymous

GMAS Subject Test Mathematics Grade 6

One–Step Equations

✎ Find the answer for each equation.

1) $3x = 90, x =$ ___

2) $5x = 35, x =$ ___

3) $6x = 24, x =$ ___

4) $24x = 144, x =$ ___

5) $x + 15 = 20, x =$ ___

6) $x - 7 = 4, x =$ ___

7) $x - 9 = 2, x =$ ___

8) $x + 15 = 23, x =$ ___

9) $x - 4 = 13, x =$ ___

10) $12 = 16 + x, x =$ ___

11) $x - 10 = 2, x =$ ___

12) $5 - x = -11, x =$ ___

13) $28 = -6 + x, x =$ ___

14) $x - 20 = -35, x =$ ___

15) $x + 14 = -4, x =$ ___

16) $14 = 28 - x, x =$ ___

17) $7 + x = -7, x =$ ___

18) $x - 16 = 4, x =$ ___

19) $30 = x - 15, x =$ ___

20) $x - 5 = -18, x =$ ___

21) $x - 10 = 24, x =$ ___

22) $x - 20 = -25, x =$ ___

23) $x - 17 = 30, x =$ ___

24) $-70 = x - 28, x =$ ___

25) $x - 9 = 13, x =$ ___

26) $36 = 4x, x =$ ___

27) $x - 35 = 25, x =$ ___

28) $x - 25 = 10, x =$ ___

29) $70 - x = 16, x =$ ___

30) $x - 10 = 14, x =$ ___

31) $17 - x = -13, x =$ ___

32) $x - 9 = -30, x =$ ___

WWW.MathNotion.Com

> GMAS Subject Test Mathematics Grade 6

One–Step Equation Word Problems

✎ Solve.

1) How many boxes of envelopes can you buy with $40 if one box costs $5?

2) After paying $8.15 for a salad, Riya has $61.53. How much money did she have before buying the salad?

3) How many packages of Tissues can you buy with $81 if one package costs $4.5?

4) Last week Joe ran 40 miles more than Harrison. Joe ran 78 miles. How many miles did Harrison run?

5) Last Friday Liam had $65.46. Over the weekend he received some money for cleaning the attic. He now has $60. How much money did he receive?

6) After paying $17.36 for a sandwich, Elise has $31.23. How much money did she have before buying the sandwich?

GMAS Subject Test Mathematics Grade 6

Two-Steps Equations

✏️ **Solve each equation.**

1) $6(3 + x) = 42$

2) $(-7)(x - 2) = 56$

3) $(-8)(3x - 4) = (-16)$

4) $5(2 + x) = -15$

5) $19(3x + 11) = 38$

6) $4(2x + 2) = 24$

7) $5(8 + 3x) = (-20)$

8) $(-5)(5x - 3) = 40$

9) $2x + 12 = 16$

10) $\dfrac{4x - 5}{5} = 3$

11) $(-3) = \dfrac{x + 4}{7}$

12) $80 = (-8)(x - 3)$

13) $\dfrac{x}{3} + 7 = 19$

14) $\dfrac{1}{4} = \dfrac{1}{2} + \dfrac{x}{4}$

15) $\dfrac{11 + x}{5} = (-6)$

16) $(-3)(10 + 5x) = (-15)$

17) $(-3x) + 12 = 24$

18) $\dfrac{x + 5}{5} = -5$

19) $\dfrac{x + 23}{8} = 3$

20) $(-4) + \dfrac{x}{2} = (-14)$

21) $-5 = \dfrac{x + 7}{8}$

22) $\dfrac{9x - 3}{6} = 4$

23) $\dfrac{2x - 12}{8} = 6$

24) $40 = (-5)(x - 8)$

GMAS Subject Test Mathematics Grade 6

Multi–Step Equations

✎ Find the answer for each equation.

1) $3x + 3 = 9$

2) $-x + 5 = 12$

3) $4x - 8 = 8$

4) $-(3 - x) = 5$

5) $4x - 8 = 16$

6) $12x - 15 = 9$

7) $2x - 18 = 2$

8) $4x + 8 = 16$

9) $24x + 27 = 75$

10) $-14(3 + x) = 14$

11) $-3(2 + x) = 6$

12) $12 = -(x - 7)$

13) $3(3 - x) = 30$

14) $-15 = -(3x + 6)$

15) $40(3 + x) = 40$

16) $5(x - 10) = 25$

17) $-18 = x + 8x$

18) $3x + 25 = -2x - 10$

19) $7(6 + 3x) = -63$

20) $18 - 3x = -4 - 5x$

21) $4 - 6x = 36 + 2x$

22) $15 + 15x = -5 + 5x$

23) $42 = (-6x) - 7 + 7$

24) $21 = 3x - 21 + 4x$

25) $-18 = -6x - 9 + 3x$

26) $5x - 15 = -29 + 6x$

27) $7x - 18 = 4x + 3$

28) $-7 - 4x = 5(4 - x)$

29) $x - 5 = -5(-3 - x)$

30) $13x - 68 = 15x - 102$

31) $-5x - 3 = -3(9 + 3x)$

32) $-2x - 15 = 6x + 17$

WWW.MathNotion.Com

One-Step Inequalities

✍ **Solve each inequality.**

1) $7x < 14$

2) $x + 7 \geq -8$

3) $x - 1 \leq 9$

4) $-2x + 4 > -10$

5) $x + 18 \geq -6$

6) $x + 9 \geq 5$

7) $x - \frac{1}{3} \leq 5$

8) $-7x < 42$

9) $-x + 8 > -3$

10) $\frac{x}{3} + 3 > -9$

11) $-x + 8 > -4$

12) $x - 14 \leq 18$

13) $-x - 5 \leq -7$

14) $x + 26 \geq -13$

15) $x + \frac{1}{3} \geq -\frac{2}{3}$

16) $x + 6 \geq -14$

17) $x - 42 \leq -48$

18) $x - 5 \leq 4$

19) $-x + 5 > -6$

20) $x + 6 \geq -12$

21) $8x + 6 \leq 22$

22) $4x - 3 \geq 9$

23) $3x - 5 < 22$

24) $6x - 8 \leq 40$

Graphing Inequalities

✎ Draw a graph for each inequality.

1) $x > -1$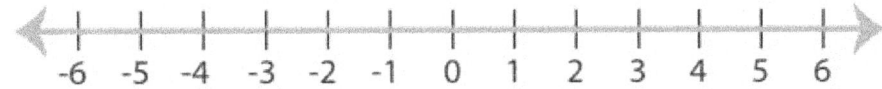

2) $x \leq 2$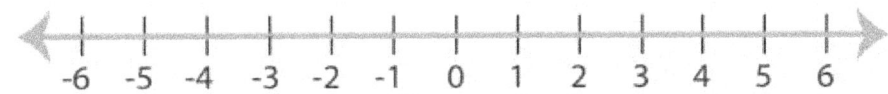

3) $x \geq 0$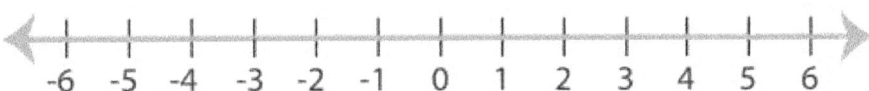

4) $x < -3$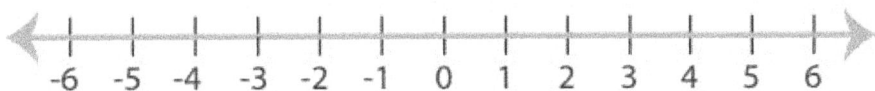

5) $x < \frac{1}{2}$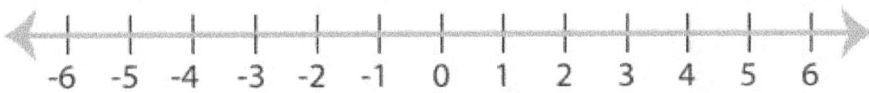

6) $x \leq -2$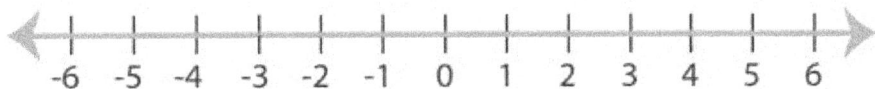

7) $x \leq 3$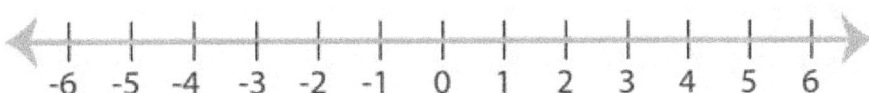

8) $x \geq -\frac{7}{2}$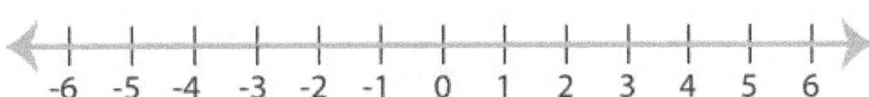

GMAS Subject Test Mathematics Grade 6

Two-Steps Inequality

✎ Solve each inequality

1) $2x - 3 \leq 7$

2) $3x - 4 \leq 8$

3) $\frac{-1}{4}x + \frac{x}{2} \leq \frac{1}{8}$

4) $5x + 10 \geq 30$

5) $4x - 7 \geq 9$

6) $3x - 5 \leq 16$

7) $8x - 2 \leq 14$

8) $9x + 5 \leq 23$

9) $2x + 10 > 32$

10) $\frac{x}{8} + 2 \leq 4$

11) $3x + 4 \geq 37$

12) $3x - 8 < 10$

13) $6 \geq \frac{x+7}{2}$

14) $3x + 9 < 48$

15) $\frac{4+x}{5} \geq 3$

16) $16 + 4x < 36$

17) $16 > 6x - 8$

18) $5 + \frac{x}{3} < 6$

19) $-4 + 4x > 24$

20) $5 + \frac{x}{7} < 3$

WWW.MathNotion.Com

Multi-Step Inequalities

✏️ **Calculate each inequality.**

21) $x - 3 \leq 7$

22) $8 - x \leq 8$

23) $3x - 9 \leq 9$

24) $4x - 4 \geq 8$

25) $x - 7 \geq 1$

26) $5x - 15 \leq 5$

27) $6x - 8 \leq 4$

28) $-11 + 6x \leq 12$

29) $4(x - 4) \leq 16$

30) $3x - 10 \leq 11$

31) $5x - 25 < 25$

32) $9x - 5 < 22$

33) $20 - 7x \geq -15$

34) $33 + 6x < 45$

35) $8 + 8x \geq 96$

36) $7 + 3x < 13$

37) $4x - 3 < 9$

38) $5(2 - 2x) \geq -30$

39) $-(7 + 6x) < 29$

40) $12 - 8x \geq -20$

41) $-4(x - 6) > 24$

42) $\dfrac{3x + 9}{6} \leq 10$

43) $\dfrac{4x - 10}{3} \leq 2$

44) $\dfrac{2x - 8}{3} > 2$

45) $8 + \dfrac{x}{6} < 9$

46) $\dfrac{9x}{7} - 4 < 5$

47) $\dfrac{15x + 45}{15} > 1$

48) $16 + \dfrac{x}{4} < 6$

GMAS Subject Test Mathematics Grade 6

Answers of Worksheets

One–Step Equations

1) 30
2) 7
3) 4
4) 6
5) 5
6) 11
7) 11
8) 8
9) 17
10) −4
11) 12
12) 16
13) 34
14) −15
15) −18
16) 14
17) −14
18) 20
19) 45
20) −13
21) 34
22) −5
23) 47
24) −42
25) 22
26) 9
27) 60
28) 35
29) 54
30) 24
31) 30
32) −21

One–Step Equation Word Problems

1) 8
2) $69.68
3) 18
4) 38
5) 5.46
6) 48.59

Two Steps Equations

1) $x = 4$
2) $x = -6$
3) $x = 2$
4) $x = -5$
5) $x = -3$
6) $x = 2$
7) $x = -4$
8) $x = -1$
9) $x = 2$
10) $x = 5$
11) $x = -25$
12) $x = -7$
13) $x = 36$
14) $x = -1$
15) $x = -41$
16) $x = -1$
17) $x = -4$
18) $x = -30$
19) $x = 1$
20) $x = -20$
21) $x = -47$
22) $x = 3$
23) $x = 30$
24) $x = 0$

Multi–Step Equations

1) 2
2) −7
3) 4
4) 8
5) 6
6) 2
7) 10
8) 2
9) 2
10) −4
11) −4
12) −5
13) −7
14) 3
15) −2
16) 15
17) −2
18) −7
19) −5
20) −11
21) −4
22) −2
23) −7
24) 6
25) 3
26) 14
27) 7
28) 27

WWW.MathNotion.Com

GMAS Subject Test Mathematics Grade 6

29) -5 30) 17 31) -6 32) -4

One Step Inequality

1) $x < 2$
2) $x \geq -15$
3) $x \leq 10$
4) $x < 7$
5) $x \geq -24$
6) $x \geq -4$
7) $x \leq \frac{16}{3}$
8) $x > -6$
9) $x < 11$
10) $x > -36$
11) $x < 12$
12) $x \leq 32$
13) $x \geq 2$
14) $x \geq -39$
15) $x \geq -1$
16) $x \geq -20$
17) $x \leq -6$
18) $x \leq 9$
19) $x < 11$
20) $x \geq -18$
21) $x \leq 2$
22) $x \geq 3$
23) $x < 9$
24) $x \leq 8$

Graphing Single–Variable Inequalities

1)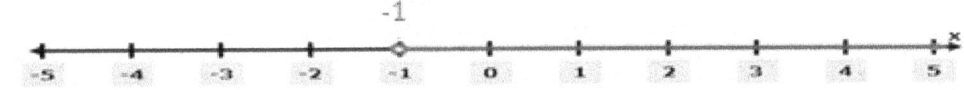

2)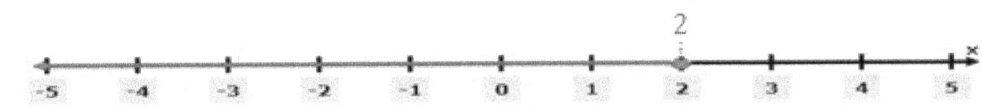

3)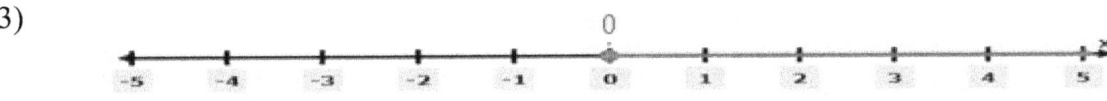

4)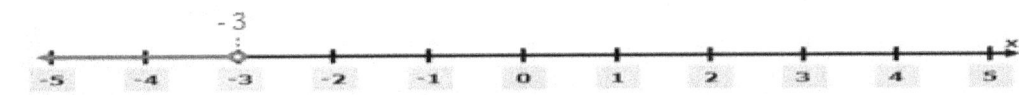

5)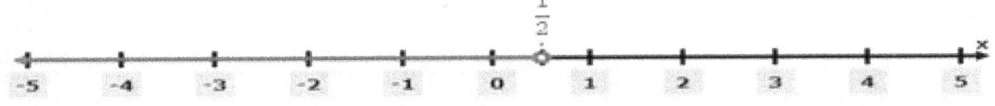

6)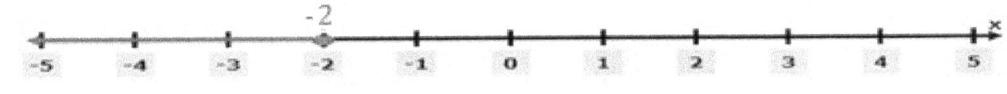

7)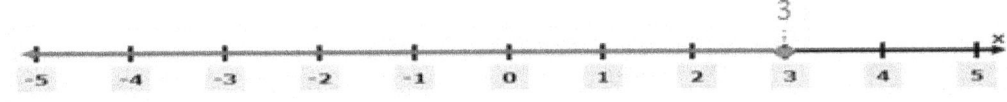

8)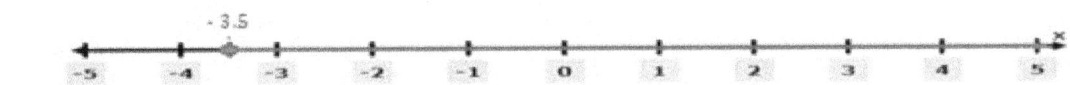

Two Steps Inequality

1) $x \leq 5$
2) $x \leq 4$
3) $x \leq 0.5$
4) $x \geq 4$
5) $x \geq 4$
6) $x \leq 7$

GMAS Subject Test Mathematics Grade 6

7) $x \leq 2$
8) $x \leq 2$
9) $x > 11$
10) $x \leq 16$
11) $x \geq 11$

12) $x < 6$
13) $x \leq 5$
14) $x < 13$
15) $x \geq 11$
16) $x < 5$

17) $x < 4$
18) $x < 3$
19) $x > 7$
20) $x < -14$

Multi-Step Inequalities

1) $x \leq 10$
2) $x \geq 0$
3) $x \leq 6$
4) $x \geq 3$
5) $x \geq 8$
6) $x \leq 4$
7) $x \leq 2$

8) $x \leq \frac{23}{6}$
9) $x \leq 8$
10) $x \leq 7$
11) $x < 10$
12) $x < 3$
13) $x \leq 5$
14) $x < 2$

15) $x \geq 11$
16) $x < 2$
17) $x < 3$
18) $x \leq 4$
19) $x > -6$
20) $x \leq 4$
21) $x < 0$

22) $x \leq 17$
23) $x \leq 4$
24) $x > 7$
25) $x < 6$
26) $x < 7$
27) $x > -2$
28) $x < -40$

GMAS Subject Test Mathematics Grade 6

Chapter 10:
Geometry and Solid Figures

Topics that you will practice in this chapter:

- ✓ Angles
- ✓ Pythagorean Relationship
- ✓ Triangles
- ✓ Polygons
- ✓ Trapezoids
- ✓ Circles
- ✓ Cubes
- ✓ Rectangular Prism
- ✓ Cylinder

Mathematics is, as it were, a sensuous logic, and relates to philosophy as do the arts, music, and plastic art to poetry. — K. Shegel

GMAS Subject Test Mathematics Grade 6

Angles

✎ **What is the value of x in the following figures?**

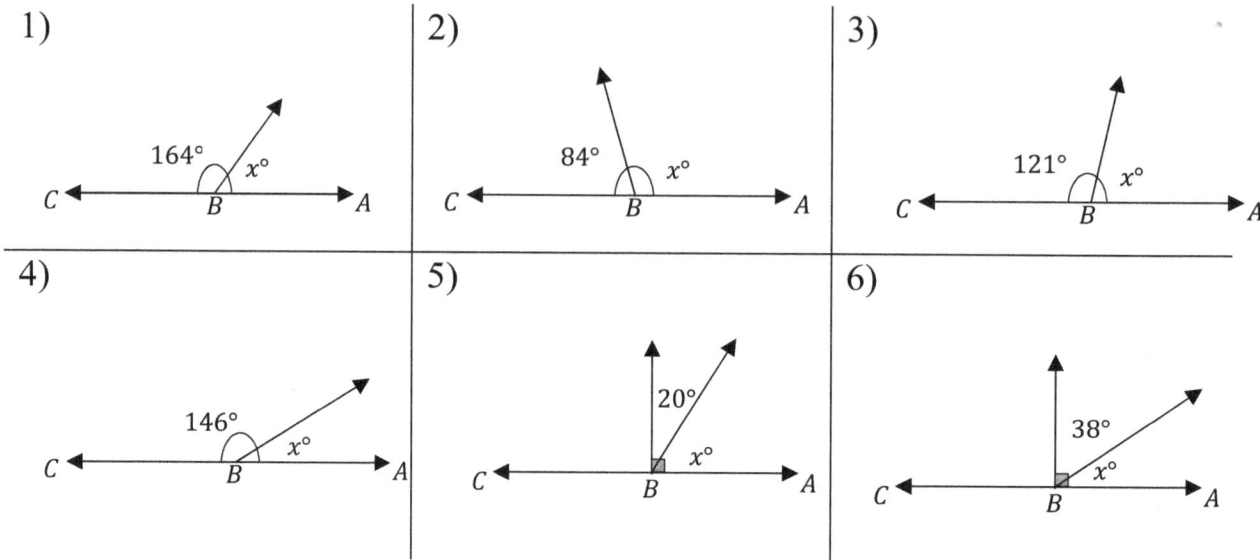

✎ **Calculate.**

7) Two supplement angles have equal measures. What is the measure of each angle? _____

8) The measure of an angle is seven fifth the measure of its supplement. What is the measure of the angle? _____

9) Two angles are complementary and the measure of one angle is 24 less than the other. What is the measure of the smaller angle? _____

10) Two angles are complementary. The measure of one angle is one fifth the measure of the other. What is the measure of the bigger angle? _____

11) Two supplementary angles are given. The measure of one angle is 40° less than the measure of the other. What does the smaller angle measure? _____

WWW.MathNotion.Com

GMAS Subject Test Mathematics Grade 6

Pythagorean Relationship

✎ Do the following lengths form a right triangle?

1)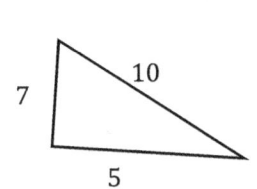

2)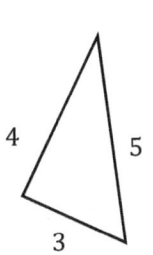

3)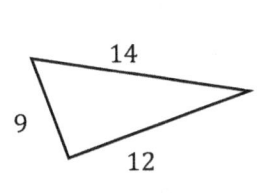

4)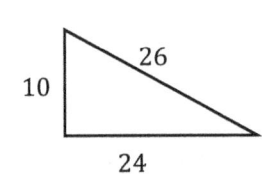

5)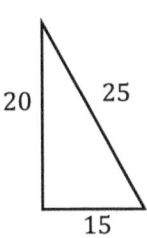

6)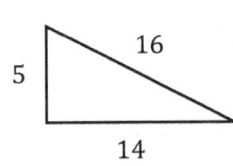

7)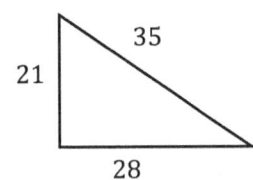

8)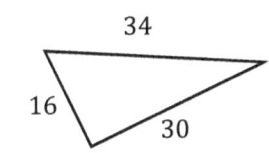

✎ Find the missing side?

9)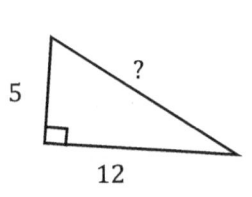

10)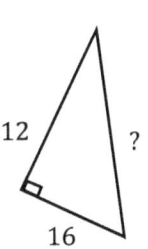

11)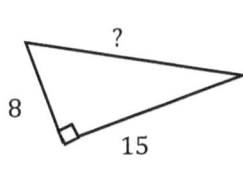

12)

13)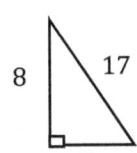

14)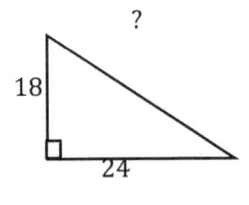

15)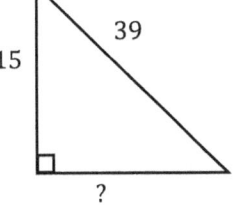

16)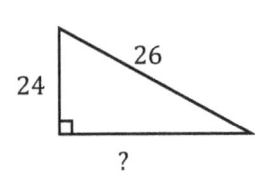

WWW.MathNotion.Com

GMAS Subject Test Mathematics Grade 6

Triangles

✎ Find the measure of the unknown angle in each triangle.

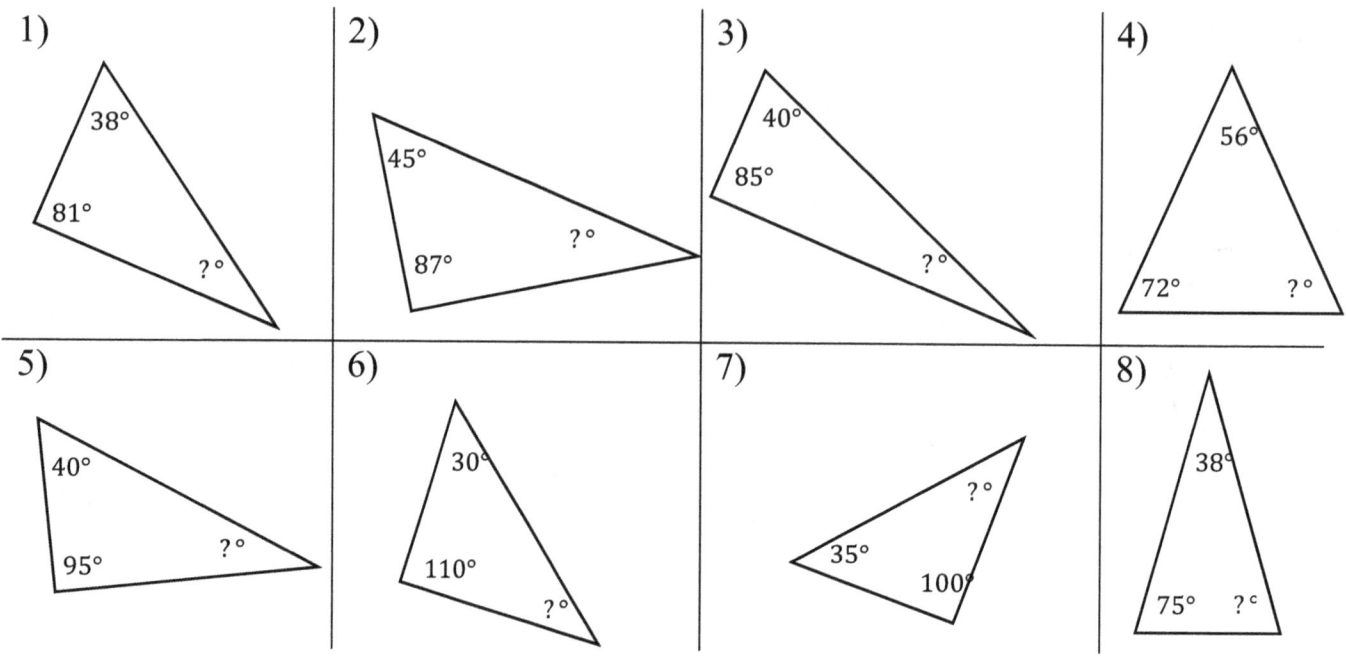

✎ Find area of each triangle.

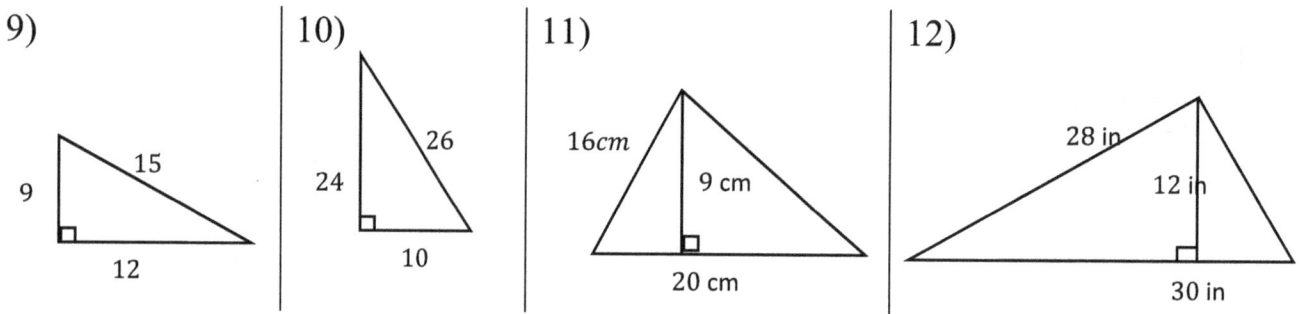

WWW.MathNotion.Com

Polygons

✏️ **Find the perimeter of each shape.**

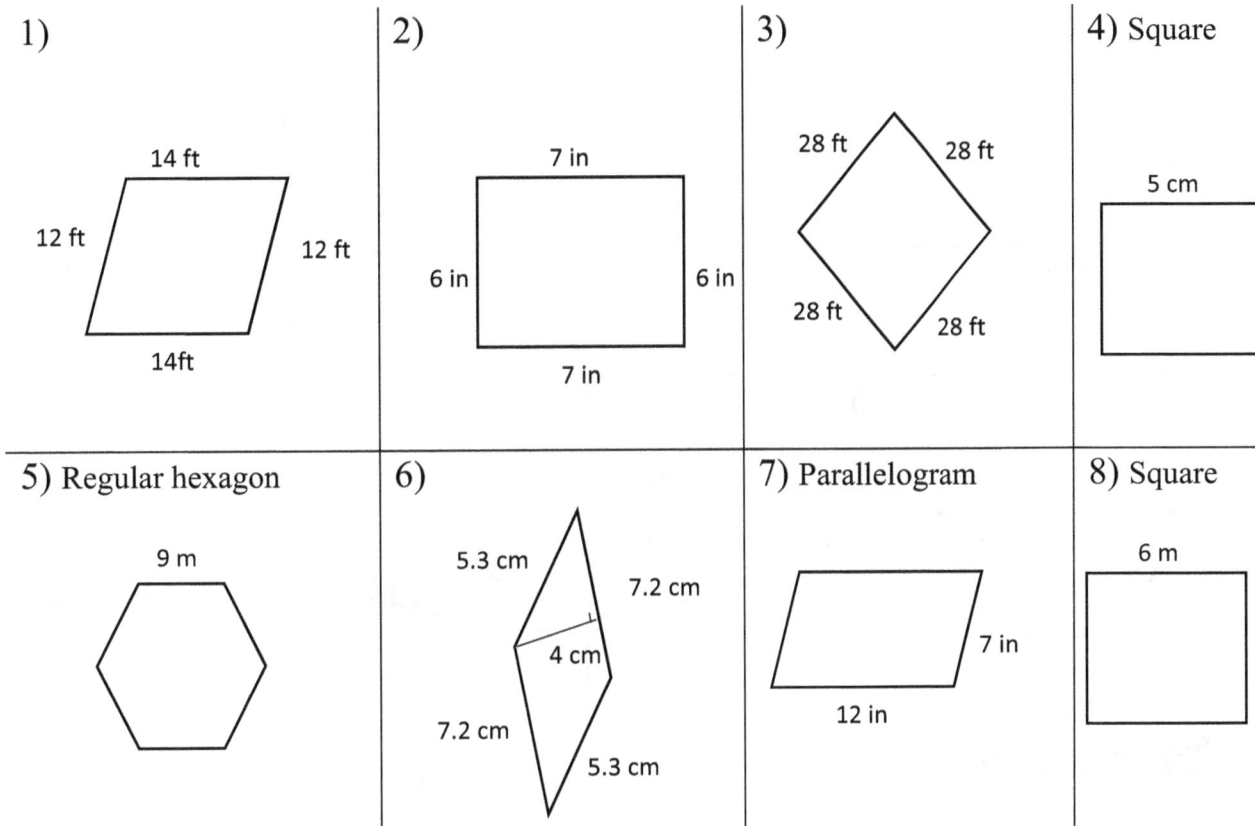

✏️ **Find the area of each shape.**

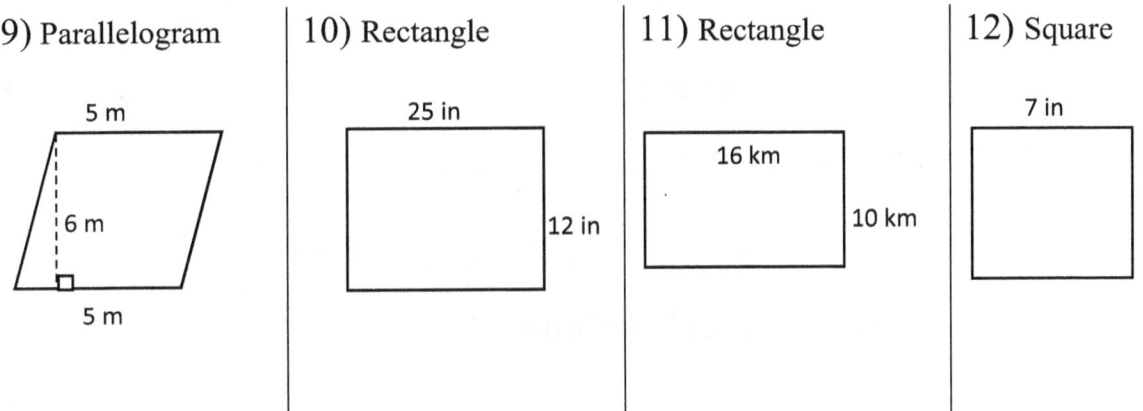

GMAS Subject Test Mathematics Grade 6

Trapezoids

✎ Find the area of each trapezoid.

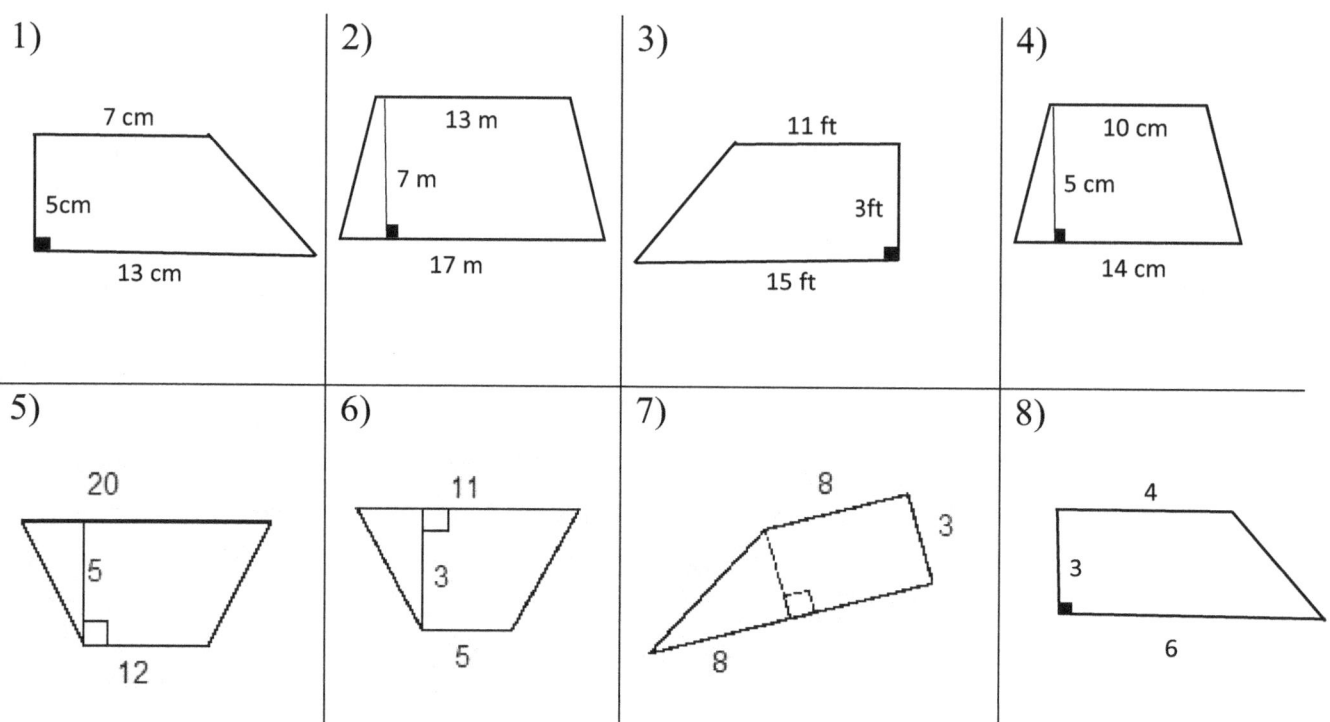

✎ Calculate.

1) A trapezoid has an area of 45 cm² and its height is 5 cm and one base is 5 cm. What is the other base length? _____

2) If a trapezoid has an area of 99 ft² and the lengths of the bases are 8 ft and 10 ft, find the height? _____

3) If a trapezoid has an area of 126 m² and its height is 14 m and one base is 6 m, find the other base length? _____

4) The area of a trapezoid is 440 ft² and its height is 22 ft. If one base of the trapezoid is 15 ft, what is the other base length? _____

GMAS Subject Test Mathematics Grade 6

Circles

✏️ **Find the area of each circle.** ($\pi = 3.14$)

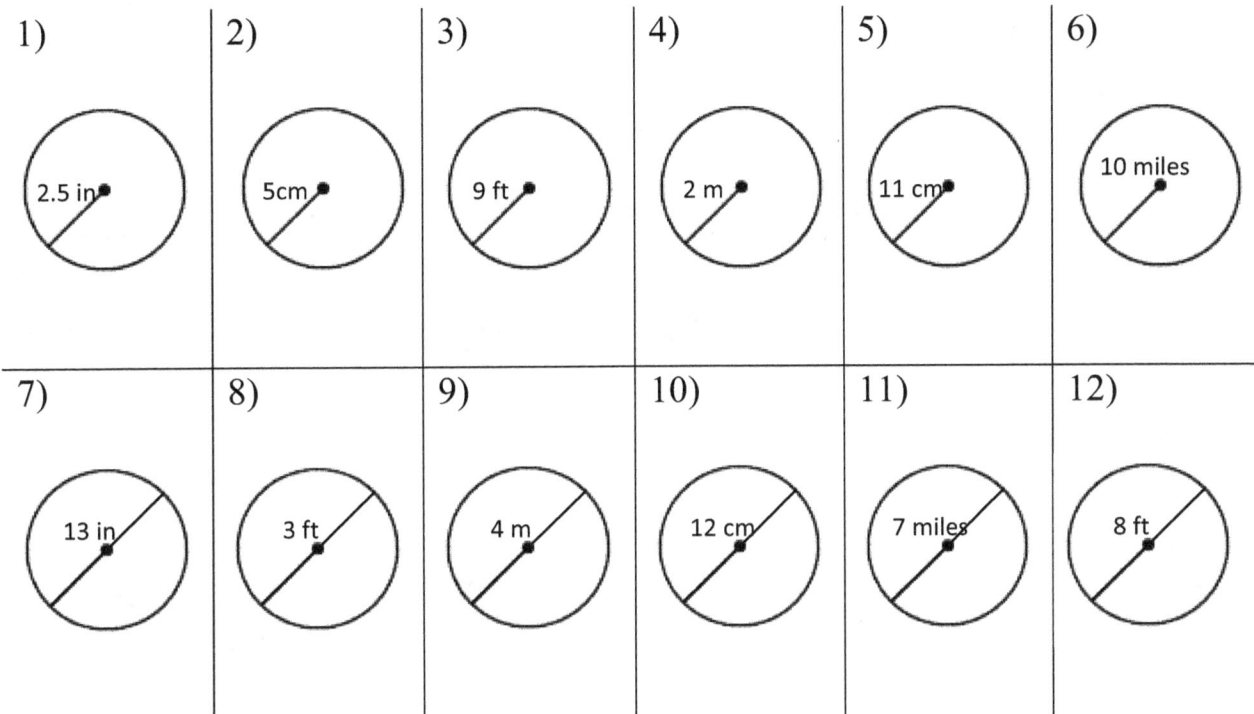

✏️ **Complete the table below.** ($\pi = 3.14$)

Circle No.	Radius	Diameter	Circumference	Area
1	1 in	2 in	6.28 in	3.14 in^2
2		10 m		
3				28.26 ft^2
4			47.1 mi	
5		11 km		
6	7 cm			
7		12 ft		
8				314 m^2
9			56.52 in	
10	4.5 ft			

WWW.MathNotion.Com

Cubes

✏ **Find the volume of each cube.**

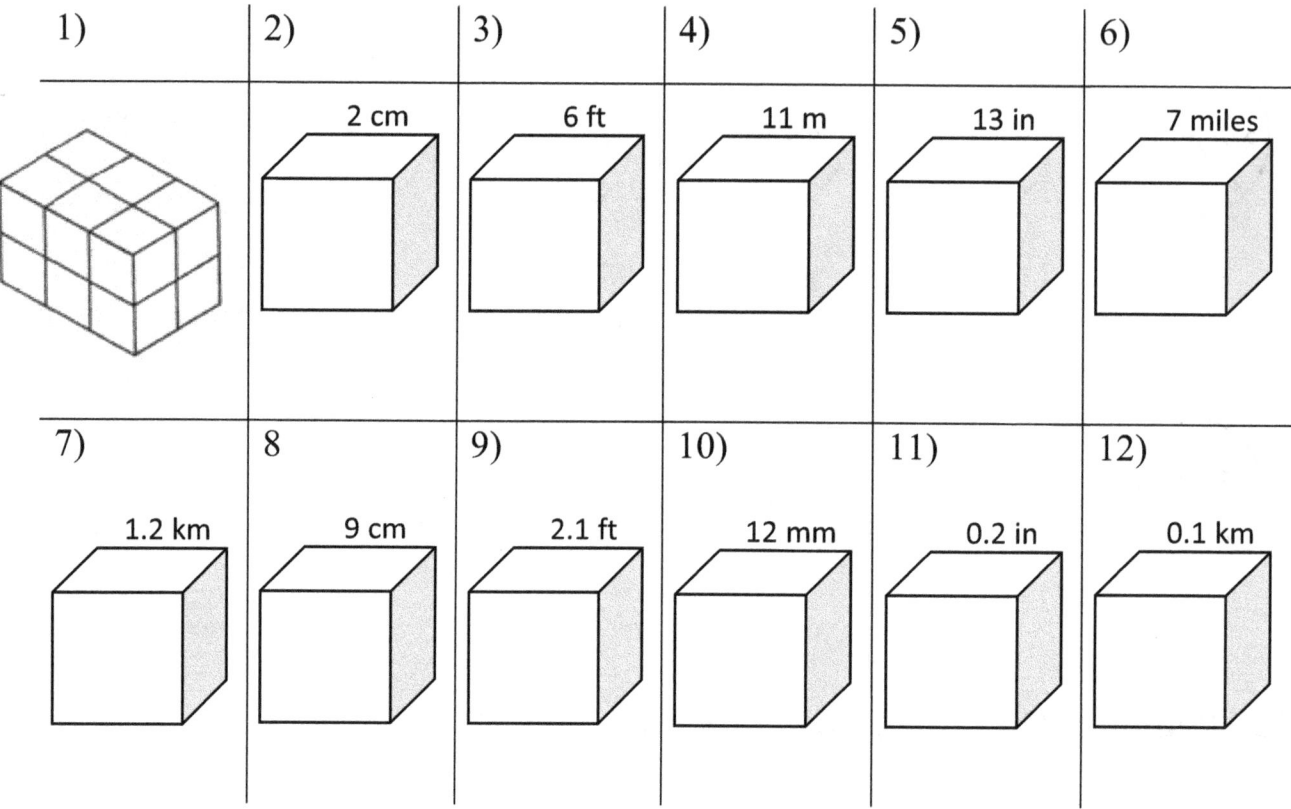

✏ **Find the surface area of each cube.**

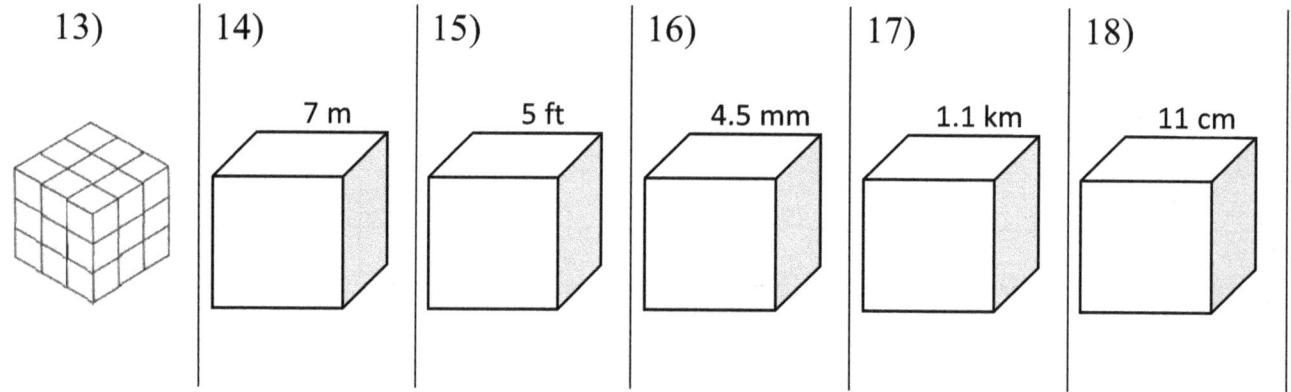

Rectangular Prism

✏️ **Find the volume of each Rectangular Prism.**

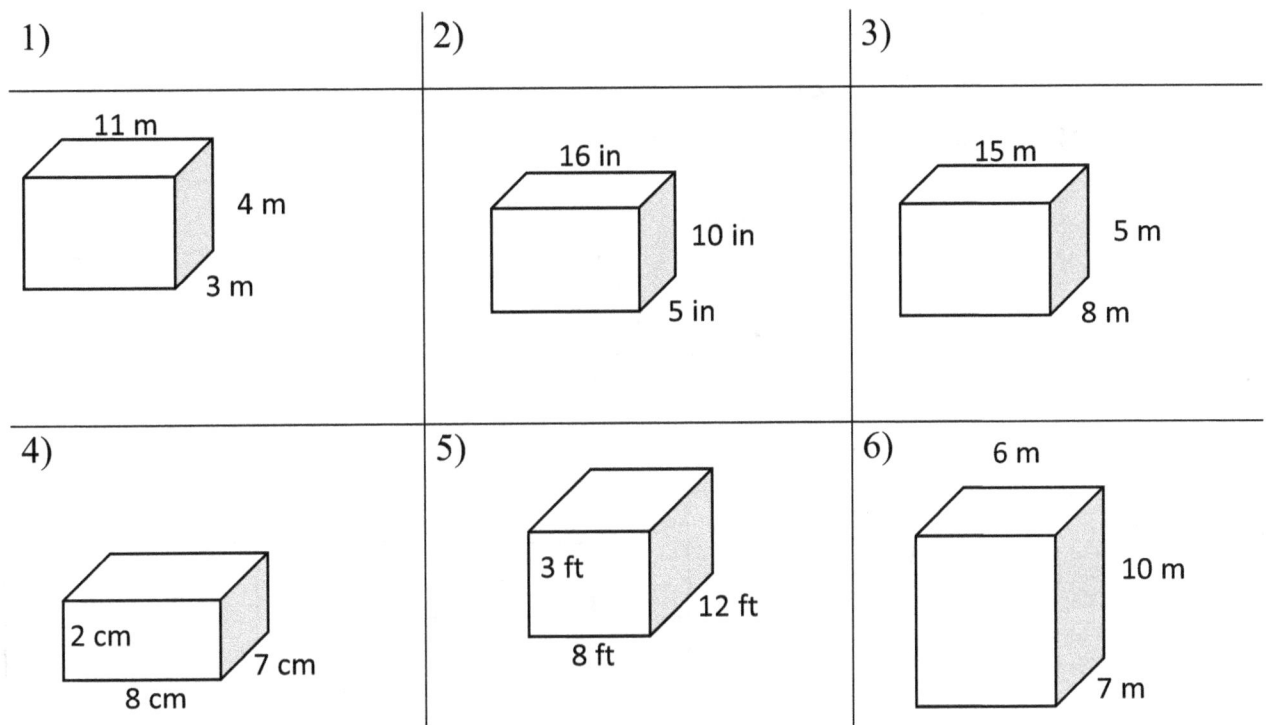

✏️ **Find the surface area of each Rectangular Prism.**

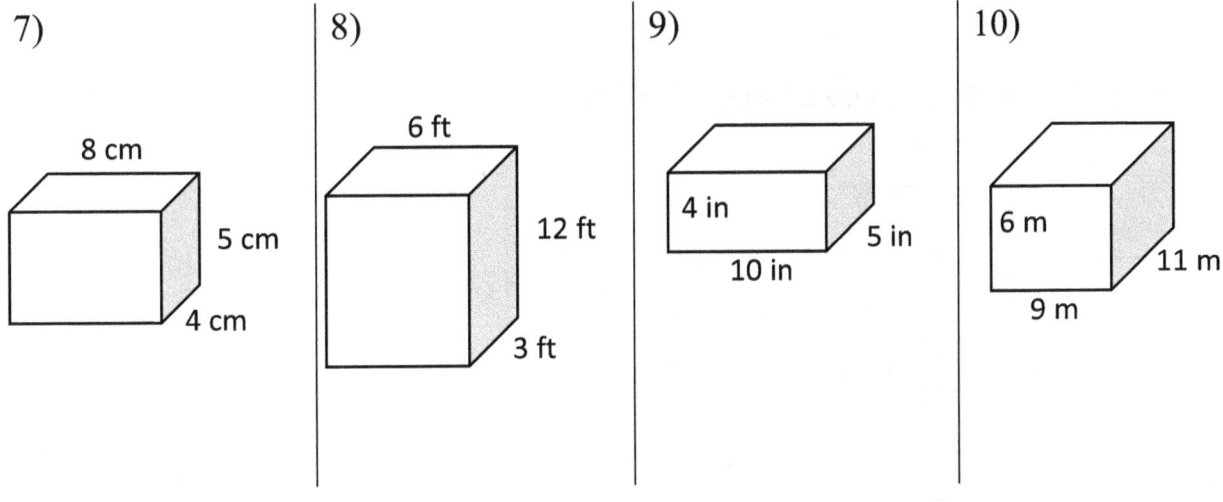

GMAS Subject Test Mathematics Grade 6

Cylinder

✏️ **Find the volume of each Cylinder. Round your answer to the nearest tenth.** ($\pi = 3.14$)

1)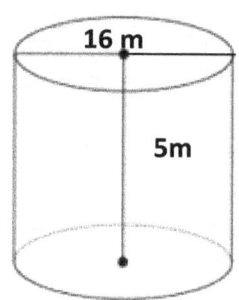

2)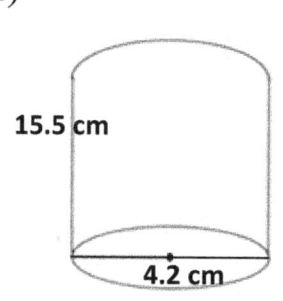

3)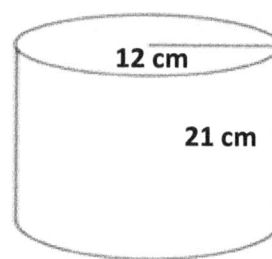

4)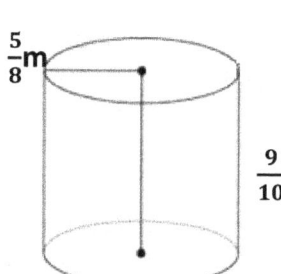

5)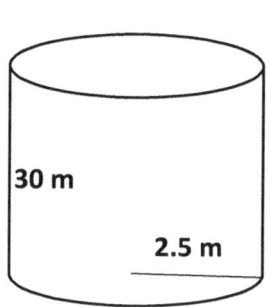

6)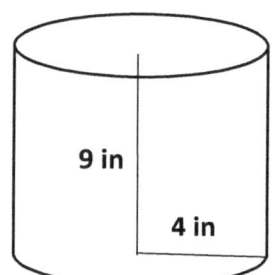

✏️ **Find the surface area of each Cylinder.** ($\pi = 3.14$)

7)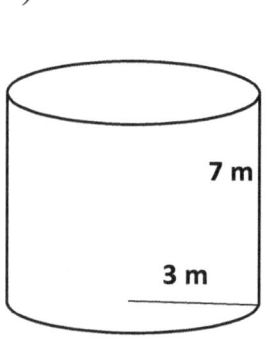

8)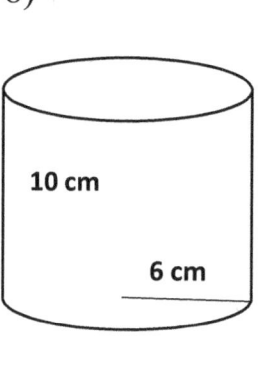

9)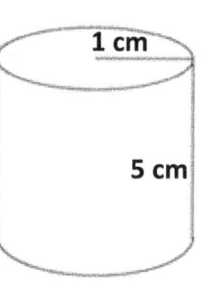

10)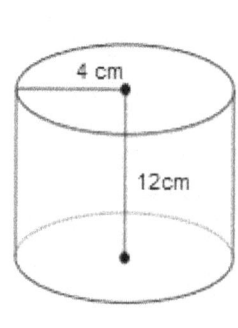

WWW.MathNotion.Com

GMAS Subject Test Mathematics Grade 6

Answers of Worksheets

Angles

1) 16°   4) 34°   7) 90°   10) 75°
2) 96°   5) 70°   8) 75°   11) 70°
3) 59°   6) 52°   9) 33°

Pythagorean Relationship

1) No   5) Yes   9) 13   13) 15
2) Yes   6) No   10) 20   14) 30
3) No   7) Yes   11) 17   15) 36
4) Yes   8) Yes   12) 10   16) 12

Triangles

1) 60°   5) 45°   9) 54 square unites
2) 48°   6) 40°   10) 120 square unites
3) 55°   7) 45°   11) 90 square unites
4) 52°   8) 67°   12) 180 square unites

Polygons

1) 52 ft   5) 54 m   9) 30 m^2
2) 26 in   6) 25 cm   10) 300 in^2
3) 112 ft   7) 38 in   11) 160 km^2
4) 20 cm   8) 24 m   12) 49 in^2

Trapezoids

1) 50 cm^2 4) 60 cm^2 7) 36
2) 105 m^2 5) 80   8) 15
3) 39 ft^2 6) 24

Calculate

1) 13 cm   2) 11 ft   3) 12 m   4) 25 ft

Circles

1) 19.63 in^2 5) 379.94 cm^2 9) 12.56 m^2
2) 78.5 cm^2 6) 314 $miles^2$ 10) 113.04 cm^2
3) 254.34 ft^2 7) 132.67 in^2 11) 38.47 $miles^2$
4) 12.56 m^2 8) 7.07 ft^2 12) 50.24 ft^2

WWW.MathNotion.Com

GMAS Subject Test Mathematics Grade 6

Circle No.	Radius	Diameter	Circumference	Area
1	1 in	2 in	6.28 in	3.14 in^2
2	5 m	10 m	31.4 m	78.5 m^2
3	3 ft	6 ft	18.84 ft	28.26 ft^2
4	7.5 miles	15 mi	47.1 mi	176.63 mi^2
5	5.5 km	11 km	34.54 km	94.99 km^2
6	7 cm	14 cm	43.96 cm	153.86 cm^2
7	6 ft	12 ft	37.68 feet	113.04 ft^2
8	10 m	20 m	62.8 m	314 m^2
9	9 in	18 in	56.52 in	254.34 in^2
10	4.5 ft	9 ft	28.26 ft	63.585 ft^2

Cubes

1) 12
2) 8 cm^3
3) 216 ft^3
4) 1,331 m^3
5) 2,197 in^3
6) 343 $miles^3$
7) 1.728 km^3
8) 729 cm^3
9) 9.261 ft^3
10) 1,728 mm^3
11) 0.008 in^3
12) 0.001 km^3
13) 27
14) 294 m^2
15) 150 ft^2
16) 121.5 mm^2
17) 7.26 km^2
18) 726 cm^2

Rectangular Prism

1) 132 m^3
2) 800 in^3
3) 600 m^3
4) 112 cm^3
5) 288 ft^3
6) 420 m^3
7) 184 cm^2
8) 252 ft^2
9) 220 in^2
10) 438 m^2

Cylinder

1) 1,004.8 m^3
2) 214.6 cm^3
3) 9,495.4 cm^3
4) 1.1 m^3
5) 588.8 m^3
6) 452.2 in^3
7) 188.4 m^2
8) 602.9 cm^2
9) 37.7 cm^2
10) 401.9 m^2

GMAS Subject Test Mathematics Grade 6

Chapter 11 :
Statistics and Probability

Topics that you will practice in this chapter:

- ✓ Mean and Median
- ✓ Mode and Range
- ✓ Times Series
- ✓ Stem-and-Leaf Plot
- ✓ Pie Graph
- ✓ Probability Problems

Mathematics is no more computation than typing is literature.
– John Allen Paulos

GMAS Subject Test Mathematics Grade 6

Mean and Median

✎ **Find Mean and Median of the Given Data.**

1) 8, 7, 14, 4, 8

2) 14, 8, 25, 19, 16, 33, 11

3) 23, 18, 15, 12, 17

4) 34, 14, 10, 15, 6, 11

5) 10, 19, 6, 8, 32, 20, 17

6) 17, 26, 39, 69, 20, 6

7) 40, 38, 18, 11, 9, 2, 7, 32, 41

8) 24, 21, 31, 12, 33, 32, 22

9) 16, 14, 20, 41, 15, 20, 38, 4

10) 20, 20, 30, 18, 6, 28, 12, 46

11) 12, 7, 10, 11, 16, 22

12) 10, 29, 27, 12, 2, 15, 10, 3

✎ **Calculate.**

13) In a javelin throw competition, five athletics score 56, 34, 62, 23 and 19 meters. What are their Mean and Median? _____

14) Eva went to shop and bought 8 apples, 14 peaches, 6 bananas, 4 pineapples and 12 melons. What are the Mean and Median of her purchase? _____

15) Bob has 17 black pen, 19 red pen, 14 green pens, 20 blue pens and 5 boxes of yellow pens. If the Mean and Median are 19 respectively, what is the number of yellow pens in each box? _____

WWW.MathNotion.Com

GMAS Subject Test Mathematics Grade 6

Mode and Range

✏️ **Find Mode and Rage of the Given Data.**

1) 4, 3, 7, 3, 3, 4
   Mode: _____   Range: _____

2) 18, 18, 24, 26, 18, 8, 14, 22
   Mode: _____   Range: _____

3) 8, 8, 8, 16, 19, 22, 20, 9, 13
   Mode: _____   Range: _____

4) 24, 24, 14, 28, 20, 18, 20, 24
   Mode: _____   Range: _____

5) 6, 21, 27, 24, 27, 27
   Mode: _____   Range: _____

6) 21, 8, 8, 7, 8, 12, 10, 22, 18, 13
   Mode: _____   Range: _____

7) 7, 4, 4, 6, 13, 13, 13, 0, 2, 2
   Mode: _____   Range: _____

8) 5, 8, 5, 14, 12, 14, 3, 5, 18
   Mode: _____   Range: _____

9) 7, 7, 7, 12, 7, 3, 8, 16, 3, 17
   Mode: _____   Range: _____

10) 15, 15, 19, 16, 4, 16, 10, 15
    Mode: _____   Range: _____

11) 6, 6, 5, 6, 42, 13, 19, 2
    Mode: _____   Range: _____

12) 8, 8, 9, 8, 9, 4, 34, 22
    Mode: _____   Range: _____

✏️ **Calculate.**

13) A stationery sold 12 pencils, 56 red pens, 24 blue pens, 20 notebooks, 12 erasers, 21 rulers and 11 color pencils. What are the Mode and Range for the stationery sells?

    Mode: _____   Range: _____

14) In an English test, eight students score 10, 15, 15, 18 18, 16, 15 and 15. What are their Mode and Range? _____

15) What is the range of the first 6 even numbers greater than 8?
    _____

WWW.MathNotion.Com

GMAS Subject Test Mathematics Grade 6

Times Series

✎ **Use the following Graph to complete the table.**

Day	Distance (km)
1	
2	

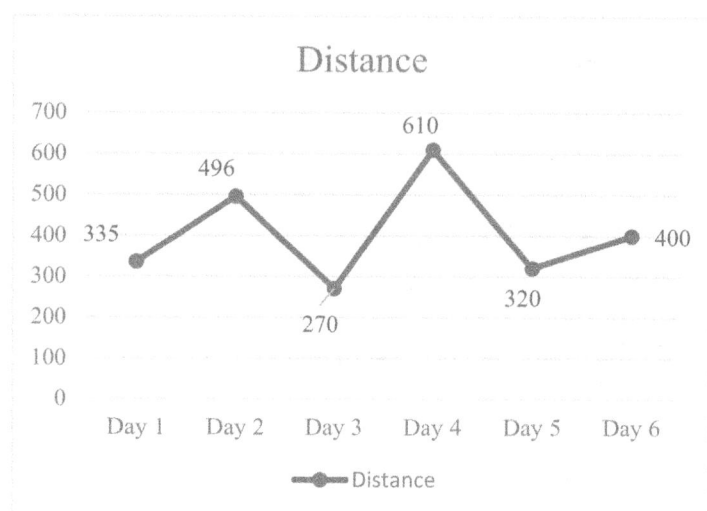

The following table shows the number of births in the US from 2007 to 2012 (in millions).

Year	Number of births (in millions)
2007	4.15
2008	3.70
2009	3.45
2010	3.20
2011	1.75
2012	2.98

Draw a Time Series for the table.

WWW.MathNotion.Com

GMAS Subject Test Mathematics Grade 6

Stem-and-Leaf Plot

✎ **Make stem ad leaf plots for the given data.**

1) 24, 26, 29, 20, 53, 27, 51, 55, 36, 21, 37, 30   Stem | Leaf plot

2) 11, 59, 66, 14, 18, 19, 59, 65, 69, 61, 68, 65   Stem | Leaf plot

3) 121, 55, 66, 54, 112, 128, 63, 125, 59, 123, 68, 119   Stem | Leaf plot

4) 51, 32, 100, 56, 84, 36, 107, 56, 85, 39, 56, 106, 89   Stem | Leaf plot

5) 33, 89, 19, 87, 81, 16, 11, 30, 86, 35, 17, 35, 13   Stem | Leaf plot

6) 60, 92, 22, 25, 67, 93, 95, 62, 21, 64, 98, 29   Stem | Leaf plot

WWW.MathNotion.Com

GMAS Subject Test Mathematics Grade 6

Quartile of a Data Set

✎ **Find First, Second and Third Quartile of the Given Data.**

1) 45, 8, 25, 43, 24, 36, 35, 62, 19

2) 23, 63, 25, 19, 80, 32

3) 86, 33, 85, 60, 72, 42, 51, 46

4) 24, 44, 48, 25, 25, 36, 25, 36, 71, 49

5) 23, 15, 45, 9, 35, 8, 25, 15

6) 66, 86, 40, 32, 82, 25, 52, 44, 61

Box and Whisker Plots

✎ **Make box and whisker plots for the given data.**

1) 86, 65, 92, 67, 72, 87, 87, 83, 95, 66, 76, 82

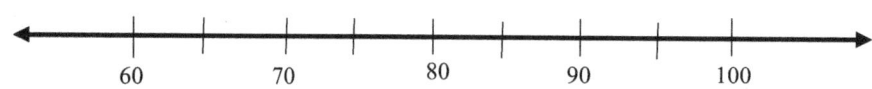

2) 8, 22, 17, 15, 13, 5, 8, 12, 6, 11, 6, 15, 4, 28

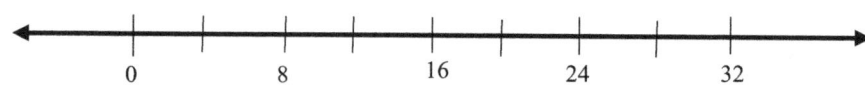

3) 25, 21, 34, 19, 23, 24, 13, 17, 15, 16, 22

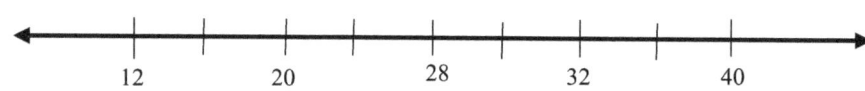

WWW.MathNotion.Com

Pie Graph

The circle graph below shows all Robert's expenses for last month. Robert spent $140 on his hobbies last month.

Answer following questions based on the Pie graph.

1) How much was Robert's total expenses last month? _____

2) How much did Robert spend on his car last month? _____

3) How much did Robert spend for shopping last month? _____

4) How much did Robert spend on his rent last month? _____

5) What fraction is Robert's expenses for his rent and car out of his total expenses last month? _____

GMAS Subject Test Mathematics Grade 6

Probability Problems

✎ **Calculate.**

1) A number is chosen at random from 1 to 10. Find the probability of selecting number 6 or smaller numbers. _____

2) Bag A contains 18 red marbles and 6 green marbles. Bag B contains 16 black marbles and 8 orange marbles. What is the probability of selecting a green marble at random from bag A? What is the probability of selecting a black marble at random from Bag B? _____

3) A number is chosen at random from 1 to 20. What is the probability of selecting multiples of 4? _____

4) A card is chosen from a well-shuffled deck of 52 cards. What is the probability that the card will be a queen? _____

5) A number is chosen at random from 1 to 15. What is the probability of selecting a multiple of 3 or 5? _____

A spinner numbered 1–8, is spun once. What is the probability of spinning …?

6) an Odd number? _____   7) a multiple of 2? _____

8) a multiple of 5? _____   9) number 10? _____

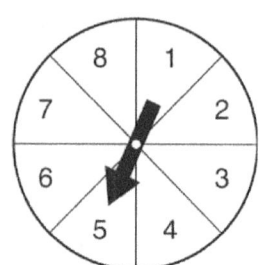

GMAS Subject Test Mathematics Grade 6

Answers of Worksheets

Mean and Median

1) Mean: 8.2, Median: 8
2) Mean: 18, Median: 16
3) Mean: 17, Median: 17
4) Mean: 15, Median: 12.5
5) Mean: 16, Median: 17
6) Mean: 29.5, Median: 23
7) Mean: 22, Median: 18
8) Mean: 25, Median: 24
9) Mean: 21, Median: 18
10) Mean: 22.5, Median: 20
11) Mean: 13, Median: 11.5
12) Mean: 13.5, Median: 11
13) Mean: 38.8, Median: 34
14) Mean: 8.8, Median: 8
15) 5

Mode and Range

1) Mode: 3, Range: 4
2) Mode: 18, Range: 18
3) Mode: 8, Range: 14
4) Mode: 24, Range: 14
5) Mode: 27, Range: 21
6) Mode: 8, Range: 15
7) Mode: 13, Range: 13
8) Mode: 5, Range: 15
9) Mode: 7, Range: 14
10) Mode: 15, Range: 15
11) Mode: 6, Range: 40
12) Mode: 8, Range: 30
13) Mode: 12, Range: 45
14) Mode: 15, Range: 8
15) 10

Time series

Day	Distance (km)
1	335
2	496
3	270
4	610
5	320
6	400

Stem–And–Leaf Plot

1)

Stem	leaf
2	0 1 4 6 7 9
3	0 6 7
5	1 3 5

2)

Stem	leaf
1	1 4 8 9
5	9 9
6	1 5 5 6 8 9

3)

Stem	leaf
5	4 5 9
6	3 6 8
11	2 9
12	1 3 5 8

GMAS Subject Test Mathematics Grade 6

4)

Stem	leaf
3	2 6 9
5	1 6 6 6
8	4 5 9
10	0 6 7

5)

Stem	leaf
1	1 3 6 7 9
3	0 3 5 5
8	1 6 7 9

6)

Stem	leaf
2	2 1 5 9
6	0 2 4 7
9	2 3 5 8

Quartile of the Given Data

1) First quartile: 21.5, second quartile: 35, third quartile: 44
2) First quartile: 22, second quartile: 28.5, third quartile: 67.25
3) First quartile: 43, second quartile: 55.5, third quartile: 81.75
4) First quartile: 25, second quartile: 36, third quartile: 48.25
5) First quartile: 10.5, second quartile: 19, third quartile: 32.5
6) First quartile: 36, second quartile: 52, third quartile: 74

Box and Whisker Plots

1)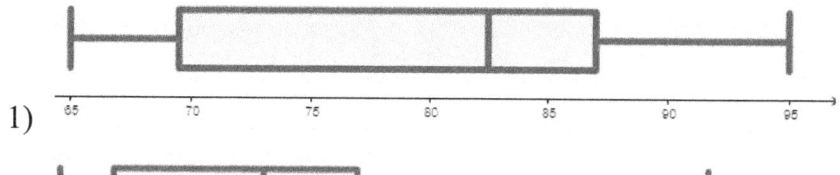

2)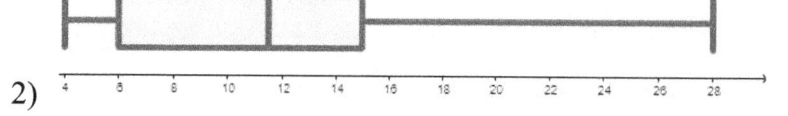

3)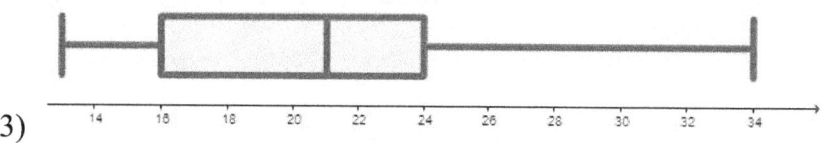

Pie Graph

1) $5,600
2) $784
3) $1,288
4) $2,016
5) $\frac{1}{2}$

Probability Problems

1) $\frac{3}{5}$
2) $\frac{1}{4}, \frac{2}{3}$
3) $\frac{1}{4}$
4) $\frac{1}{13}$
5) $\frac{7}{15}$
6) $\frac{1}{2}$
7) $\frac{1}{2}$
8) $\frac{1}{8}$
9) 0

GMAS Subject Test Mathematics Grade 6

Chapter 12 : GMAS Math Practice Tests

Time to Test

Time to refine your skill with a practice examination.

Take a REAL GMAS Mathematics test to simulate the test day experience. After you've finished, score your test using the answer key.

Before You Start

- You'll need a pencil and scratch papers to take the test.
- For this practice test, don't time yourself. Spend time as much as you need.
- It's okay to guess. You won't lose any points if you're wrong.
- After you've finished the test, review the answer key to see where you went wrong.

Calculators are not permitted for GMAS Tests.

Good Luck!

GMAS Subject Test Mathematics Grade 6

GMAS Grade 6 Mathematics Reference Sheet

Conversions:

LENGTH

Customary	Metric
1 mile (mi) = 1,760 yards (yd)	1 kilometer (km) = 1,000 meters (m)
1 yard (yd) = 3 feet (ft)	1 meter (m) = 100 centimeters (cm)
1 foot (ft) = 12 inches (in.)	1 centimeter (cm) = 10 millimeters (mm)

VOLUME AND CAPACITY

Customary	Metric
1 gallon (gal) = 4 quarts (qt)	1 liter (L) = 1,000 milliliters (mL)
1 quart (qt) = 2 pints (pt.)	
1 pint (pt.) = 2 cups (c)	
1 cup (c) = 8 fluid ounces (Fl oz)	

WEIGHT AND MASS

Customary	Metric
1 ton (T) = 2,000 pounds (lb.)	1 kilogram (kg) = 1,000 grams (g)
1 pound (lb.) = 16 ounces (oz)	1 gram (g) = 1,000 milligrams (mg)

Formulas:

Area

Triangle	$A = \frac{1}{2}bh$
Rectangle or Parallelogram	$A = bh$
Trapezoid	$A = \frac{1}{2}h(b_1 + b_2)$

Volume

Rectangular Prism	$V = Bh$

GMAS Subject Test Mathematics Grade 6

Georgia Milestones Assessment System Practice Test 1

Mathematics

GRADE 6

Released Month Year

GMAS Subject Test Mathematics Grade 6

Session 1

- ❖ Calculators are permitted for this practice test.
- ❖ Time for Session 1: 85 Minutes

GMAS Subject Test Mathematics Grade 6

1) If $x = -4$, which of the following equations is true?

   A. $x(4x + 8) = 38$

   B. $4(27 - x^2) = -42$

   C. $3(-4x - 6) = 35$

   D. $x(5x + 7) = 52$

2) What is the perimeter of the following shape? (it's a right triangle)

   A. 28 cm

   B. 15 cm

   C. 18 cm

   D. 36 cm

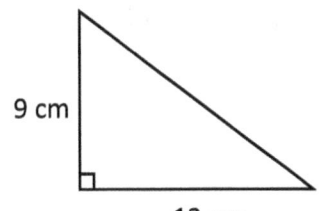

3) 40 is what percent of 16?

   Write your answer in the box below.

4) Which of the following expressions has a value of -16?

   A. $-7 + (-21 \div 3) + \frac{-3}{8} \times 8$

   B. $3 \times (-9) + (-4) \times 5$

   C. $(-8) + 14 \times 4 \div (-7)$

   D. $(-5) \times (-3) + 6$

GMAS Subject Test Mathematics Grade 6

5) A football team won exactly 60% of the games it played during last session. Which of the following could be the total number of games the team played last season?

   A. 24

   B. 45

   C. 38

   D. 32

6) 7 less than quadruple a positive integer is 61. What is the integer?

   A. 19

   B. 17

   C. 15

   D. 21

7) Which of the following expressions has the greatest value?

   A. $3^5 - 8^2$

   B. $15^2 - 6^3$

   C. $5^3 - 4^3$

   D. $2^7 - 9^2$

GMAS Subject Test Mathematics Grade 6

8) The diameter of a circle is 7π. What is the area of the circle?

A. $\dfrac{49\pi^3}{4}$

B. $\dfrac{49\pi^2}{2}$

C. $\dfrac{14\pi^3}{4}$

D. $49\pi^3$

9) Elise has x apples. Alvin has 45 apples, which is 30 apples less than number of apples Elise owns. If Baron has $\dfrac{1}{5}$ times as many apples as Elise has. How many apples does Baron have?

A. 13

B. 17

C. 15

D. 22

10) Find the perimeter of shape in the following figure? (all angles are right angles)

Write your answer in the box below.

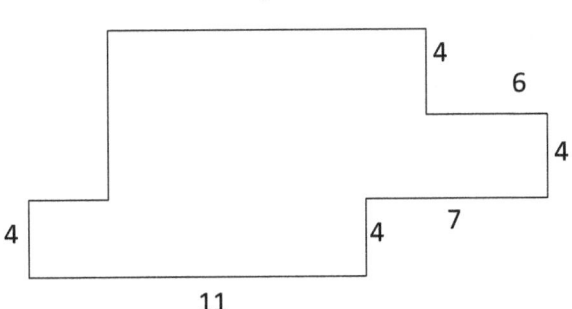

GMAS Subject Test Mathematics Grade 6

Session 2

❖ Calculators are permitted for this practice test.

❖ Time for Session 2: 85 Minutes

GMAS Subject Test Mathematics Grade 6

11) Car A travels 180.40 km at a given time, while car B travels 1.8 times the distance car A travels at the same time. What is the distance car B travels during that time?

   A. 223.37 km

   B. 433.27 km

   C. 334.27 km

   D. 324.72 km

12) What is the probability of choosing a month starts with F in a year?

   A. $\frac{5}{12}$

   B. $\frac{1}{3}$

   C. $\frac{1}{4}$

   D. $\frac{1}{12}$

13) What are the values of mode and median in the following set of numbers?

   1, 3, 2, 2, 5, 4, 9, 3, 7, 2, 4

   A. Mode: 2, Median: 2.5

   B. Mode: 2, 3 Median: 4

   C. Mode: 2, Median: 2

   D. Mode: 2, Median: 3

GMAS Subject Test Mathematics Grade 6

14) If point A placed at $-\frac{36}{9}$ on a number line, which of the following points has a distance equal to 10 from point A?

   A. -14

   B. 5

   C. 6

   D. A and C

15) The ratio of pens to pencils in a box is 3 to 4. If there are 84 pens and pencils in the box altogether, how many more pens should be put in the box to make the ratio of pens to pencils 1: 1?

   Write your answer in the box below.

16) Which of the following shows the numbers in increasing order?

   A. $\frac{17}{3}, \frac{9}{4}, \frac{5}{6}, \frac{42}{5}$

   B. $\frac{9}{4}, \frac{5}{6}, \frac{17}{3}, \frac{42}{5}$

   C. $\frac{5}{6}, \frac{9}{4}, \frac{17}{3}, \frac{42}{5}$

   D. $\frac{42}{5}, \frac{9}{4}, \frac{5}{6}, \frac{17}{3}$

17) If $8x - 3 = 29$, what is the value of $4x + 4$?

   A. 16

   B. 20

   C. 12

   D. 30

18) $5(1.25) - 4.50 = \cdots ?$

Write your answer in the box below.

19) In the following triangle find α.

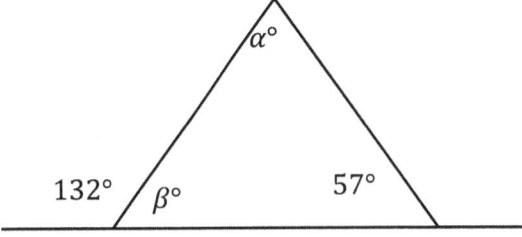

Write your answer in the box below.

20) A shaft rotates 260 times in 8 seconds. How many times does it rotate in 10 seconds?

A. 245

B. 325

C. 250

D. 340

Practice Test 1

This is the End of this Section

Georgia Milestones Assessment System

Practice Test 2

Mathematics

GRADE 6

Released

Session 1

- ❖ Calculators are permitted for this practice test.
- ❖ Time for Session 1: 85 Minutes

GMAS Subject Test Mathematics Grade 6

1) Martin earns $60 an hour. Which of the following inequalities represents the amount of time Martin needs to work per day to earn at least $630 per day?

   A. $60t \geq 630$

   B. $60t \leq 630$

   C. $60 + t \geq 630$

   D. $60 + t \leq 630$

2) What is the value of the expression $5(2x - 3y) + (9 - 4x)^2$, when $x = 2$ and $y = -1$?

   Write your answer in the box below.

   ☐

3) Round $\frac{241}{9}$ to the nearest tenth.

   A. 32.5

   B. 25

   C. 26.8

   D. 30.5

4) Which expression is equivalent to $7(11x - 5)$?

   A. -35

   B. $-77x$

   C. $77x - 35$

   D. $35x - 70$

GMAS Subject Test Mathematics Grade 6

5) A chemical solution contains 6% alcohol. If there is 24 ml of alcohol, what is the volume of the solution?

   A. 420 ml

   B. 440 ml

   C. 400 ml

   D. 1,240 ml

6) Which ordered pair describes point A that is shown below?

   A. $(5, -1)$

   B. $(-5, 1)$

   C. $(-5, -1)$

   D. $(5, 1)$

   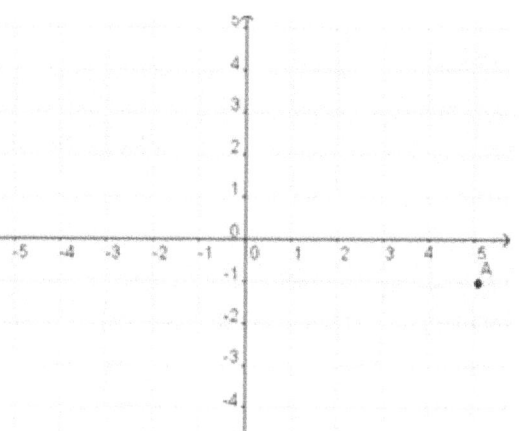

7) To produce a special concrete, for every 20 kg of cement, 4 liters of water is required. Which of the following ratios is the same as the ratio of cement to liters of water?

   A. 80: 16

   B. 75: 12

   C. 40: 16

   D. 60: 20

GMAS Subject Test Mathematics Grade 6

8) Find the opposite of the numbers 12, 6.

Write your answer in the box below.

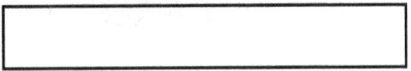

9) What is the value of x in the following equation?

$$-28 = 53 - x$$

A. 81

B. −81

C. 95

D. −95

10) Which of the following graphs represents the following inequality?

$$2 \leq 3x - 7 < 17$$

A.

B.

C.

D.

Session 2

- ❖ Calculators are permitted for this practice test.
- ❖ Time for Session 2: 85 Minutes

GMAS Subject Test Mathematics Grade 6

11) The ratio of boys to girls in a school is 3:7. If there are 340 students in the school, how many boys are in the school?

   A. 218

   B. 238

   C. 120

   D. 102

12) $(53 + 7) \div 15$ is equivalent to ...

   A. $60 \div 1.5$

   B. $\frac{60}{15} + 15$

   C. $(2 \times 2 \times 3 \times 5) \div (3 \times 5)$

   D. $(2 \times 3 \times 5 \times 5) \div 3 + 5$

13) What is the equation of a line that passes through points (4, 6) and (5, 8)?

   A. $y = -2x$

   B. $y = 2x - 2$

   C. $y = 2x + 2$

   D. $y = x + 2$

GMAS Subject Test Mathematics Grade 6

14) What is the volume of a box with the following dimensions? Height = 5 cm

Width = 7 cm   Length = 9 cm

Write your answer in the box below.

☐

15) Anita's trick–or–treat bag contains 18 pieces of chocolate, 22 suckers, 27 pieces of gum, 21 pieces of licorice. If she randomly pulls a piece of candy from her bag, what is the probability of her pulling out a piece of sucker?

A. $\frac{1}{5}$

B. $\frac{1}{4}$

C. $\frac{7}{18}$

D. $\frac{1}{22}$

16) What is the lowest common multiple of 21 and 75?

A. 525

B. 63

C. 255

D. 275

17) Which statement is true about all square?

A. Both diagonals have equal measure.

B. All sides are congruent.

C. Both diagonals are perpendicular.

D. All the statements are true.

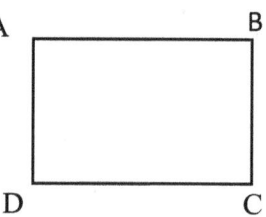

WWW.MathNotion.Com

GMAS Subject Test Mathematics Grade 6

18) Which of the following lists shows the fractions in order from least to greatest?

$$\frac{8}{9}, \frac{3}{7}, \frac{27}{8}, \frac{15}{13}$$

A. $\frac{15}{13}, \frac{8}{9}, \frac{3}{7}, \frac{27}{8}$

B. $\frac{8}{9}, \frac{15}{13}, \frac{27}{8}, \frac{3}{7}$

C. $\frac{3}{7}, \frac{8}{9}, \frac{15}{13}, \frac{27}{8}$

D. $\frac{15}{13}, \frac{8}{9}, \frac{27}{8}, \frac{3}{7}$

19) Which statement about 4 multiplied by $\frac{3}{5}$ must be true?

A. The product is between 3 and 4.

B. The product is greater than 2.8.

C. The product is equal to $\frac{30}{14}$.

D. The product is between 1.8 and 2.8.

20) If the area of the following rectangular ABCD is 100, and E is the midpoint of AB, what is the area of the shaded part?

Write your answer in the box below.

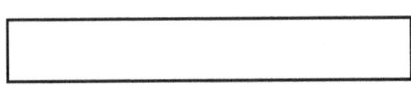

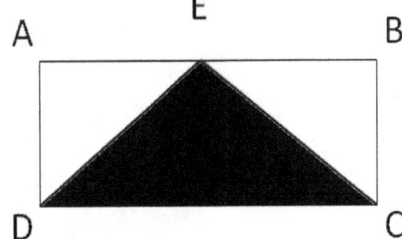

Practice Test 2
This is the End of this Section.

GMAS Subject Test Mathematics Grade 6

Chapter 13 : Answers and Explanations

Georgia Milestones Assessment System Practice Tests Answer Key

❉ Now, it's time to review your results to see where you went wrong and what areas you need to improve!

Practice Test - 1

1	D	11	D
2	D	12	D
3	250%	13	D
4	C	14	D
5	B	15	22
6	B	16	C
7	A	17	B
8	A	18	1.75
9	C	19	75°
10	60	20	B

Practice Test - 2

1	A	11	D
2	36	12	C
3	C	13	B
4	C	14	315
5	C	15	B
6	A	16	A
7	A	17	D
8	−12, −6	18	C
9	A	19	D
10	C	20	50

GMAS Subject Test Mathematics Grade 6

Practice Test 1
Georgia Milestones Assessment System - Mathematics
Answers and Explanations

1) Answer: B

Plugin the value of x in the equations. $x = -4$, then:

A. $x(4x + 8) = 38 \rightarrow -4(4(-4) + 8) = -4(-16 + 8) = -4(-8) = 32 \neq 38$

B. $4(27 - x^2) = -42 \rightarrow 4(27 - (-4)^2) = 4(27 - 16) = 4(11) = 44 \neq -42$

C. $3(-4x - 6) = 35 \rightarrow 3(-4(-4) - 6) = 3(16 - 6) = 3(10) = 30 \neq 35$

D. $x(5x + 7) = 52 \rightarrow -4(5(-4) + 7) = -4(-20 + 7) = -4(-13) = 52 = 52$

2) Answer: D

Use Pythagorean theorem to find the hypotenuse of the triangle.

$a^2 + b^2 = c^2 \rightarrow 9^2 + 12^2 = c^2 \rightarrow 81 + 144 = c^2 \rightarrow 225 = c^2 \rightarrow c = 15$

The perimeter of the triangle is: $9 + 12 + 15 = 36$

3) Answer: 250%

$40 = \frac{\text{percent}}{100} \times 16 \Rightarrow 40 = \frac{\text{percent} \times 16}{100} \Rightarrow$

$40 = \frac{\text{percent} \times 4}{25}$, multiply both sides by 25.

$1,000 = \text{percent} \times 4$, divide both sides by 4.

$250 = \text{percent}$; The answer is 250%

4) Answer: C

Let's check the options provided.

A. $-7 + (-21 \div 3) + \frac{-3}{8} \times 8 \rightarrow -7 + (-7) + (-3) = -17$

B. $3 \times (-9) + (-4) \times 5 = (-27) + (-20) = -47$

C. $(-8) + 14 \times 4 \div (-7) = -8 + 56 \div (-7) = -8 - 8 = -16$

D. $(-5) \times (-3) + 6 = 15 + 6 = 21$

GMAS Subject Test Mathematics Grade 6

5) Answer: B

Choices A, C and D are incorrect because 60% of each of the numbers is a non-whole number.

A. 24, $60\% \text{ of } 24 = 0.60 \times 24 = 14.4$

B. 45, $60\% \text{ of } 45 = 0.60 \times 45 = 27$

C. 38, $60\% \text{ of } 38 = 0.60 \times 38 = 22.8$

D. 32, $60\% \text{ of } 32 = 0.60 \times 32 = 19.2$

6) Answer: B

Let x be the integer. Then: $4x - 7 = 61$

Add 7 both sides: $4x = 68$

Divide both sides by 4: $x = 17$

7) Answer: A

A. $3^5 - 8^2 = 243 - 64 = 179$

B. $15^2 - 6^3 = 225 - 216 = 9$

C. $5^3 - 4^3 = 125 - 64 = 61$

D. $2^7 - 9^2 = 128 - 81 = 47$

8) Answer: A

The radius of the circle is: $\frac{7\pi}{2}$

The area of circle: $\pi r^2 = \pi(\frac{7\pi}{2})^2 = \pi \times \frac{49\pi^2}{4} = \frac{49\pi^3}{4}$

9) Answer: C

Elise has x apple which is 30 apples more than number of apples Alvin owns. Therefore:

$x - 30 = 45 \rightarrow x = 45 + 30 = 75$

Elise has 75 apples.

Let y be the number of apples that Baron has. Then: $y = \frac{1}{5} \times 75 = 15$

10) Answer: 60.

Let x and y be two sides of the shape. Then:

GMAS Subject Test Mathematics Grade 6

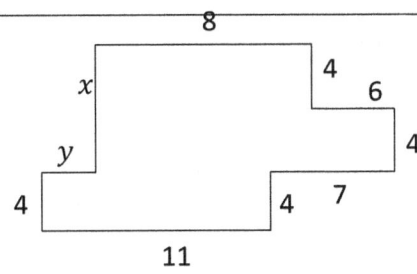

$x + 4 = 4 + 4 + 4 \rightarrow x = 8$

$y + 8 + 6 = 11 + 7 \rightarrow y + 14 = 18 \rightarrow y = 4$

Then, the perimeter is:

$4 + 11 + 4 + 7 + 4 + 6 + 4 + 8 + 8 + 4 = 60$

11) Answer: D

Distance that car B travels $= 1.8 \times$ distance that car A travels

$= 1.8 \times 180.40 = 324.72$ Km

12) Answer: D

One month, February in 12 months start with F, then:

Probability $= \dfrac{number\ of\ desired\ outcomes}{number\ of\ total\ outcomes} = \dfrac{1}{12}$

13) Answer: D

First, put the numbers in order from least to greatest: 1, 2, 2, 2, 3, 3, 4, 4, 5, 7, 9

The Mode of the set of numbers is: 2 (the most frequent numbers)

Median is: 3 (the number in the middle)

14) Answer: D

If the value of point A is greater than the value of point C, then the distance of two points on the number line is: value of A− value of C

A. $-\dfrac{36}{9} - (-14) = -4 + 14 = 10 = 10$

B. $5 - \left(-\dfrac{36}{9}\right) = 5 + 4 = 9 \neq 10$

C. $6 - \left(-\dfrac{36}{9}\right) = 6 + 4 = 10 = 10$

15) Answer: 22

$3 + 4 = 7 \Rightarrow 84 \div 7 = 12$

$84 \div 7 = 12 \rightarrow 12 \times 3 = 36 \Rightarrow 84 - 36 = 48 \Rightarrow 48 - 26 = 22$

The ratio of pens to pencils is 3: 4. Therefore there are 3 pens out of all 7 pens and pencils. To find the answer, first dived 84 by 7 then multiply the result by 3.

There are 36 pens and 48 pencils (84−36). Therefore, 22 more pens should be put in the box to make the ratio 1: 1.

GMAS Subject Test Mathematics Grade 6

16) Answer: C

$\frac{42}{5} \cong 8.4$ $\frac{9}{4} = 2.25$ $\frac{5}{6} \cong 0.83$ $\frac{17}{3} \cong 5.67$

Then: $\frac{5}{6} < \frac{9}{4} < \frac{17}{3} < \frac{42}{5}$

17) Answer: B

$8x - 3 = 29 \rightarrow 8x = 29 + 3 = 32 \rightarrow x = \frac{32}{8} = 4$

Then, $4x + 4 = 4(4) + 4 = 16 + 4 = 20$

18) Answer: 1.75

$5(1.25) - 4.50 = 6.25 - 4.5 = 1.75$

19) Answer: 75°

Supplementary angles add up to 180 degrees.

$\beta + 132° = 180° \rightarrow \beta = 180° - 132° = 48°$

The sum of all angles in a triangle is 180 degrees. Then:

$\alpha + \beta + 57° = 180° \rightarrow \alpha + 48° + 57° = 180°$

$\rightarrow \alpha + 105° = 180° \rightarrow \alpha = 180° - 105° = 75°$

20) Answer: B

The shaft rotates 260 times in 8 seconds. Then, the number of rotates in 10 second equals to:

$\frac{260 \times 10}{8} = 325$

GMAS Subject Test Mathematics Grade 6

Practice Test 2
Georgia Milestones Assessment
System - Mathematics
Answers and Explanations

1) Answer: A

For one hour he earns $60, then for t hours he earns $60t$. If he wants to earn at least $630, therefor, the number of working hours multiplied by 60 must be equal to 630 or more than 630. $60t \geq 630$

2) Answer: 36

Plug in the value of x and y and use order of operations rule.

$x = 2$ and $y = -1$

$5(2x - 3y) + (9 - 4x)^2 = 5(2(2) - 3(-1)) + (9 - 4(2))^2 = 5(4 + 3) + (9 - 8)^2$

$= 35 + 1 = 36$

3) Answer: C

$\frac{241}{9} \cong 26.78 \cong 26.8$

4) Answer: C

$7(11x - 5) = (7 \times 11x) - (7 \times 5) = (7 \times 11)x - (7 \times 5) = 77x - 35$

5) Answer: C

6% of the volume of the solution is alcohol. Let x be the volume of the solution.

Then: 6% of $x = 24$ ml $\Rightarrow 0.06\, x = 24 \Rightarrow x = 24 \div 0.06 = 400$

6) Answer: A

The coordinate plane has two axes. The vertical line is called the y-axis and the horizontal is called the x-axis. The points on the coordinate plane are address using the form (x, y). The point A is five unit on the right side of x-axis; therefore, its x value is 5 and it is one unit down, therefore its y axis is -1. The coordinate of the point is: $(5, -1)$

WWW.MathNotion.Com

GMAS Subject Test Mathematics Grade 6

7) Answer: A

$80 : 16 = 20 : 4$

$20 \times 4 = 80$ And $4 \times 4 = 16$

8) Answer: $-12, -6$

Opposite number of any number x is a number that if added to x, the result is 0. Then:

$12 + (-12) = 0$ and $6 + (-6) = 0$

9) Answer: A

$-28 = 53 - x$

First, subtract 53 from both sides of the equation. Then:

$-28 - 53 = 53 - 53 - x \to -81 = -x$, Multiply both sides by $(-1) \to x = 81$

10) Answer: C

Solve for x. $2 \leq 3x - 7 < 17 \Rightarrow$ (add 7 all sides) $2 + 7 \leq 3x - 7 + 7 < 17 + 7$

$\Rightarrow 9 \leq 3x < 24 \Rightarrow$ (divide all sides by 3) $3 \leq x < 8$

x is between 3 and 8. Choice C represent this inequality.

11) Answer: D

The ratio of boy to girls is 3:7. Therefore, there are 3 boys out of 10 students. To find the answer, first divide the total number of students by 10, then multiply the result by 3.

$340 \div 10 = 34 \Rightarrow 34 \times 3 = 102$

12) Answer: C

$(53 + 7) \div (15) = (60) \div (15)$

The prime factorization of 60 is: $2 \times 2 \times 3 \times 5$

The prime factorization of 15 is: 3×5

Therefore: $(60) \div (15) = (2 \times 2 \times 3 \times 5) \div (3 \times 5)$

13) Answer: B

The slope of the line is: $\frac{y_2 - y_1}{x_2 - x_1} = \frac{8-6}{5-4} = \frac{2}{1} = 2$

The equation of a line can be written as:

$y - y_0 = m(x - x_0) \to y - 6 = 2(x - 4) \to y - 6 = 2x - 8 \to y = 2x - 2$

WWW.MathNotion.Com

GMAS Subject Test Mathematics Grade 6

14) Answer: 315

Volume of a box = length × width × height = $9 \times 7 \times 5 = 315$

15) Answer: B

Probability = $\frac{number\ of\ desired\ outcomes}{number\ of\ total\ outcomes} = \frac{22}{18+22+27+21} = \frac{22}{88} = \frac{1}{4}$

16) Answer: A

Prime factorizing of $21 = 3 \times 7$

Prime factorizing of $75 = 3 \times 5 \times 5$

LCM = $3 \times 5 \times 5 \times 7 = 525$

17) Answer: D

In any square, all the statements are true.

18) Answer: C

Let's compare each fraction: $\frac{3}{7} < \frac{8}{9} < \frac{15}{13} < \frac{27}{8}$

Only choice C provides the right order.

19) Answer: D

$4 \times \frac{3}{5} = \frac{12}{5} = 2.4$

A. $2.4 < 3$

B. $2.4 < 2.8$

C. $\frac{30}{14} = 2.14 \neq 2.4$

D. $1.8 < 2.4 < 2.8$ This is the answer!

20) Answer: 50

Since, E is the midpoint of AB, then the area of all triangles DAE, DEF, CFE and CBE are equal.

Let x be the area of one of the triangles, then:

$4x = 100 \rightarrow x = 25$

The area of DEC = $2x = 2(25) = 50$

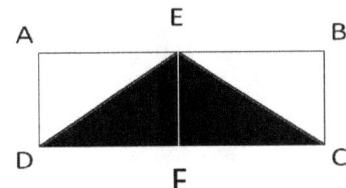

"End"